To Fuk, Susan & Co.!

With all my
love
Peter
q
Beauty or Beast?.

BEAUTY AND THE BEAST

THE AESTHETIC MOMENT IN SCIENCE

BEAUTY AND THE BEAST

THE AESTHETIC MOMENT IN SCIENCE

Ernst Peter Fischer

Translated from German by
Elizabeth Oehlkers

PLENUM TRADE • NEW YORK AND LONDON

Library of Congress Cataloging-in-Publication Data

Fischer, Ernst Peter, 1947-
    [Schöne und das Biest. English]
    Beauty and the beast : the aesthetic moment in science / Ernst
Peter Fischer ; translated from German by Elizabeth Oehlkers.
        p.   cm.
    Includes bibliographical references and index.
    ISBN 0-306-46011-4
    1. Science--Aesthetics.   I. Title.
Q172.F546913 1999
501--dc21   99-17913
                                                                  CIP

The original version of this volume was published in German under the title *Das Schöne und das Biest: Ästhetische Momente in der Wissenschaft*, © 1997 Piper Verlag GmbH, München, Germany.

ISBN 0-306-46011-4

Plenum Trade is a Division of Plenum Publishing Corporation
233 Spring Street, New York, N.Y. 10013

10 9 8 7 6 5 4 3 2 1

A C.I.P. record for this book is available from the Library of Congress

Printed in the United States of America

For Martin Rabe
and all those who make Holzen possible

Contents

Preface

Beginning with a Fairy Tale: Beauty and the Beast

The final product of nature's continuous improvements is the beautiful human being.[1]

Johann Wolfgang von Goethe

Beauty and the Beast. A musical. A film. A beautiful girl and a monster. Beauty? Beauty and science? Where's the connection? Didn't Martin Heidegger already rule this out when he assured us that "there's no such thing as beauty in the sciences"?[2] Do technological output and technical jargon equal beauty? And what kind of "aesthetic moments" go into them? Doesn't science have more to do with the quantifiable, the verifiable, and the applicable than with beauty?

There are those who think they can do without science, keep their distance from it, or hide from it, but those of us living in Western society today cannot live without it. Science is an adventure of the mind and is far more necessary for mental and spiritual edification than for the gratification of physical needs, even if it doesn't always seem that way. The effects of scientific research surround us and make their way inside us. With this kind of powerful influence, shouldn't we call for a science that demonstrates the beauty and the sensory enjoyment that accompanies its processes?

Science isn't all to blame for its current less-than-beautiful state. Currently, as in the past, science is trying hard to please people and to make their lives easier. As it has spent hundreds of years of trying to do just this, how can we say now that science has therefore lost all beauty? Beauty has held a distinguished place in science for centuries through its presentation in public alone.

Today, in fact, many consider the face of science ugly the moment it arrives with any flair or in bright colors. This is a deficiency on our part. Even in the Age of Enlightenment—a good 200 years ago—the universal genius Erasmus Darwin circulated the collective knowledge of his time in verse form (*The Temple of Nature*), and the German biologist Ernst Haeckel expressed his enthusiasm almost 100 years ago for the discoveries of natural science in his famous *Kunstformen der Natur*[3] (Fig. 1).

The fine arts and science—they once went together, and people directed the affectionate feelings traditionally reserved for successful aesthetic creations toward scientific research, even toward work meant primarily for practical application. The public no longer affectively appreciates science for its aesthetic value, and in my opinion such a deficit is harmful and creates difficulties in the long run. The fairy tale helps clarify this essential relationship by revealing the beauty of the beast.

A Few Qualifiers

When it comes to discussions on the subject of this book, it won't be long before someone, doubting, asks for a closer definition of the central metaphor—above all for that of beauty. The skeptic might say, "If you have no answers to the questions, 'What is science to you?' and 'What is beauty to you?' then there's no way we can continue our discussion." However, exactly the opposite is true. For as soon as someone actually tries to define the terms more precisely, the conversation grinds to a halt or a senseless debate ensues regarding the fine details of the definitions, and soon the actual subject and its inner tension—the connection between beauty and science—is no longer the topic of discussion.

It's not easy to formulate what makes up "the beast" nor is it clear-cut what "beauty" is. When I refer to science, however, I mean anything that would be published by serious journals under the given heading and what would be accepted by interested readers under this definition. Additionally, we can enter into a discussion of beauty if the taste for what is "beautiful," both natural and artificial, is accessible to all people without requiring special mental acrobatics.

Throughout the book I continue to present attempts at definitions, both my own and others', with those for beauty outnumbering those for

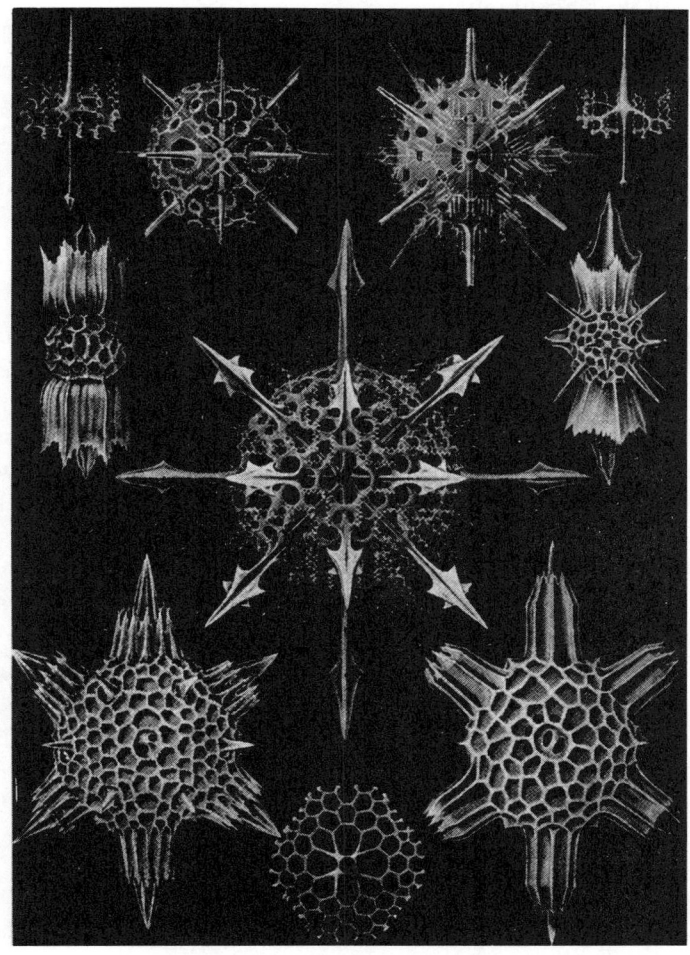

Figure 1. A diagram from the *Kunstformen der Natur* by Ernst Haeckel, which illustrates the pleasure that beauty in scientific research can generate. Different species of radiolarian impress with their symmetry—a property analyzed and appraised in its own chapter of this book.

science. In general, however, I would prefer not to focus on qualifying concepts. I am much more interested in the human activity that, even before the concepts, endeavors to conceive. By this I mean the ability to perceive, or *aisthesis* in Greek.

The link between perception and aesthetics is a critical one. The idea of aesthetics, which in today's usage has only a passing connection to scientific knowledge, is derived from a root word that indicates the reverse. The aesthetic moment in science has everything to do with this relationship of aesthetics to perception.

La Belle et la Bête

The four words, *beauty and the beast*, have various sources; one is the title of the famous film by Jean Cocteau from 1945–46, called *La Belle et la Bête* with Jean Marais in the leading role. Cocteau based his work on a fairy tale—originally by Jean Marie Leprince de Beaumont—and because his film doesn't stick very close to the storyline, to find the actual source of Beauty's tale, we have to go beyond the works I've mentioned and keep in mind the original myth that gives an account of the relationship between the beast and the beautiful girl.[4] This particular myth concerns a father with four daughters whose youngest daughter is the favorite, as anyone can imagine. (No mother ever appears in the story, a detail open to certain psychological interpretations. More interesting to me is the number 4, which will continue to reemerge throughout this book.)

When the father decides to grant his daughters a special wish, the eldest three ask for the usual, that is, for precious gifts, whereas the fourth daughter is interested only in a white rose. It sounds simple but in the end the task proves daunting because in order to get such a rose, the father must forge into the enchanted garden that grows in front of a castle owned by a wild animal.

Everything comes to pass as it must, meaning that the father is caught in the act. He envisions the worst, but to his surprise the beast proves generous. The father is permitted to first return home, unfortunately without being told whether or not he is allowed to bring the white rose with him. In any case, he is obliged to return to the garden after 3 months to receive his punishment (which was never named and therefore remains unknown).

The fourth daughter convinces her father that she should be allowed to return to the animal in place of her father, and thus the story continues. Once again the beast proves generous. He not only accepts the exchange of persons, but also imposes no punishment. On the contrary, he offers the girl a room in his castle where it's obvious she'll be able to live well. The beautiful girl does live well; she accepts the world and things as they are to such an extent that when the animal asks if they might be married one day, she does not get upset.

Only after suddenly glimpsing the face of her father in the mirror and realizing at once that he is ill does the girl become upset. She asks the animal for permission to spend a week at home to take care of her father. The beast agrees, but not without vowing his love for the girl and professing repeatedly that if she does not return in seven days, he will die.

After she returns home, however, the beautiful girl forgets all about her promise, until one day she dreams that the animal is dying of despair in his castle. At this very moment she returns to the castle to help the beast. In helping him, not only does she completely forget his ugliness, but the beautiful girl is also overcome with the feeling that the animal absolutely cannot live without her. She promises to be his wife if only he won't die.

As she comes to this decision, it grows light (both in the myth and in Cocteau's film). Music sounds, and the beast is changed into a prince (a handsome one of course). He then tells the story of a witch who enchanted him long ago. The power that trapped him in the form of a wild animal could only be broken by a beautiful girl who would selflessly love the unlucky man on account of his goodness—not his goods.

And the Moral of the Story ...

So ends the myth. Literature is teeming with psychological, psycho-analytic, and other interpretations of this myth, and I am not going to attempt to add to this already imposing mass of ingenious accomplishments. However, by linking science and the wild animal, the "beast," I can now take a look at where this comparison takes us.

If I apply the moral of the story, which could be interpreted as: Everything turns out for the best when there are people who love and respect the beast; to science, the moral of my story is that science should be appreciated not only for its utility, but also for its quality—its inner goodness. Science wanes and dies when no one is concerned with its intellectual and spiritual requirements, or when only its ugly, technical exterior receives all the attention; when individuals rob its creations and pass them off as personal achievements; or when no one revisits past works, maybe even brave failures that have lain dormant.

Science, the beast in the fairy tale, has an outwardly terrifying and malevolent presence. But like the beast, it acts generously at every turn, and it trusts those who have cared enough to get to know it and it has helped them make a home. Science is worth taking care of, in view of the myth's application to our present-day standpoint, even when it sometimes comes across as a brute. If only because it exists, science deserves our attention.

It helps to understand what the Viennese engineer and social reformer Josef Popper* wrote, who under the name of "Lynkeus" became a well-known writer at the beginning of the twentieth century. As a scientist he concerned himself with power transmissions, and in 1901 he tried to demonstrate that "technical progress" should also be valued for its "aesthetic and cultural significance." Popper wrote:

> They say that scientific progress and especially progress in the natural sciences is a dictate of reason and that progress in technology is a dictate of our utility and our comfort, but I am convinced that this only addresses one facet of the issue; I am certain that both science and technology gratify our perception in the same way art has been doing all along.[5]

---

*Josef Popper (1838–1921) was related neither by blood nor by marriage to the well-known Sir Karl Popper (1902–1994) who, as a philosopher, entered science somewhat later.

1   *Aesthetic Science: Beautiful Ideas and Elegant Experiments*

The supreme beauty of human nature exists in the light of science. [1]
St. Thomas Aquinas

There is an idea concerning the very beginnings of modern science that is often dismissed by postmodern scientists as methodologically unfounded: In the sixteenth century when Nicolaus Copernicus removed the earth from the center of the universe to make room for the sun, he didn't take this revolutionary step because it was the critical observation that testified against the old (geocentric) view in favor of the new (heliocentric) view of the heavens. Scientific progress was the least of his concerns; his primary motivation was aesthetic. In his central work of 1543 where he discusses the revolutions of celestial spheres (*De Revolutionibus Orbium Coelestium*), he grows rapturous about the idea of the sun "sitting in the middle" and implores his readers, "In the middle of all sits Sun enthroned. In this most beautiful temple could we place this luminary in any better position from which he can illuminate the whole at once?" [2]

Copernicus—as Martin Luther before him and others who, for various reasons, set their sights on the heavens—was becoming increasingly dissatisfied with the profusion of inaccurate ideas put forth by astronomers in his day. However, in spite of reports to the contrary that persist to the present, Copernicus's heliocentric design of the universe changed almost nothing. The scientific community had to wait a long time for the measurements and observations needed to defend the new, Copernican model in which the earth made a circuit around the central position of the sun and to refute the older Ptolemaic system where the earth was the stationary center of the cosmos. For purely quantitive and technical reasons, they would have to wait until the nineteenth century (see Fig. 1.1).

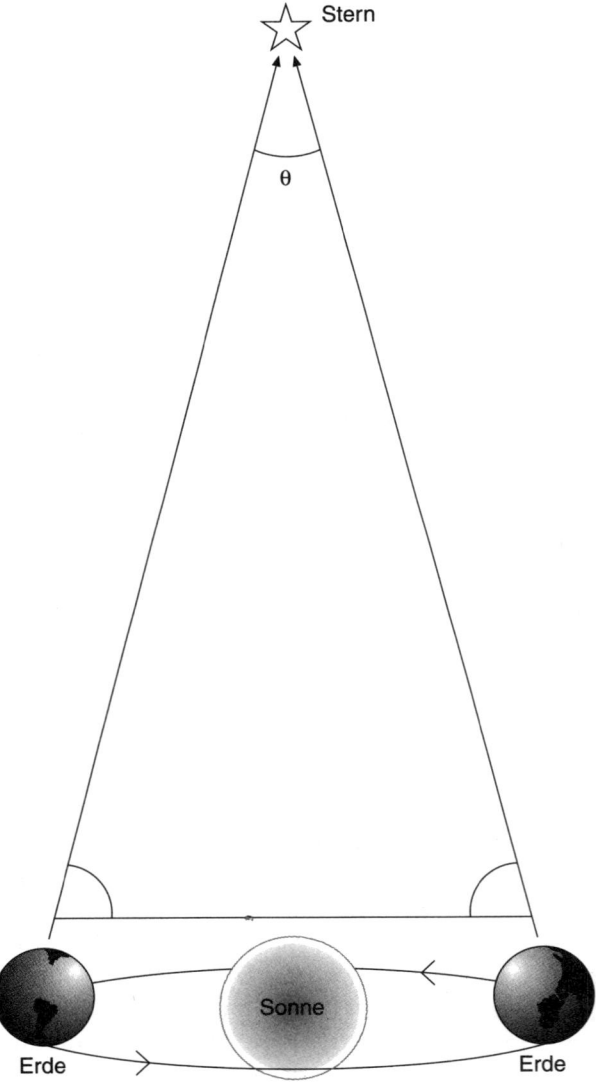

Figure 1.1.   For the heliocentric view of life to represent the celestial proportions correctly (and not just beautifully), the earth must revolve around the sun. Because it will have a different position in winter than in summer, an (arbitrary) fixed star has to be observed from the different perspectives during each season, and the parallax—an angle in between—must be measured. Unless this angle is observed, the earth's motion around the sun cannot be established. This was the argument during classical antiquity that banished expression of Aristarchus's early thoughts of a heliocentric universe. Aristarchus based his argument on an aesthetic moment:

We have, however, never let lack of empirical evidence regarding developments in cosmology interfere with our image of the universe. For most people, the availability of infallible evidence is irrelevant. This indifference to inaccuracies or technical shortcomings makes it that much clearer that we are more motivated, then as now, by the all-encompassing human need to find beauty in the universe. Copernicus shared this need. "So we find underlying this ordination an admirable symmetry in the Universe, and a clear bond of harmony in the motion and magnitude of the Spheres such as can be discovered in no other wise."[3]

BEAUTIE IS SUCH A THING AS WE COMMONLY PREFERRE BEFORE ALL THINGS
John Lyly from *Euphues and His Ephoebus*

The aesthetic element this heliocentric view of the world offered must have been at once readily apparent and generally attractive. Otherwise, how are we to understand the acceptance—without a shred of empirical evidence—of the system Copernicus proposed or the increasingly popular viewpoint that its truth was absolute? The same aesthetic element must still be at once readily apparent and attractive because very often this heliocentric view continues to hold authority even though its flaws have been recognized for some time.

Discoveries by astronomers make clearer every day just how "wrong" the heliocentric view is. We now know that the sun is not in the center of the cosmos and that it is not, as Copernicus thought, stationary. He was right only in that the earth moves much more than was previously thought and that it orbits around the sun along with many other planets. The entire solar system, including the earth and its human inhabitants, is receding

←——————————————————————————————————

he was convinced that the sun was substantially bigger than the earth, so it seemed reasonable to put it in the center. However, he couldn't establish the angle, and it didn't occur to him that the distance of the fixed star could be so great that the parallax would be too small to measure without some help (it's smaller than a second of arc). To put it in perspective: the earth's orbit is only light-minutes away; the distance of the fixed star is millions of light-years. The necessary technology for this measurement arrived only in the nineteenth century: In 1838–39 they succeeded in measuring the parallax of Star 61 Cygni and could establish the motion of the earth. That the earth was in motion, however, had been accepted and known for some time already. But what made them so certain? A. Earth; B. Sun; C. Earth; D. Star.

according to a complex formula as just a tiny part of a galaxy called the Milky Way that sits at the very edge of the universe, and with the aid of modern instruments can finally be investigated. (I shall say only in passing that some contemporary intellectuals who have problems with the traditional image of God see this marginal position of our galaxy as right and proper, even beautiful. In any case it seems to suit most of us just fine, as there haven't been too many recent complaints about its extreme proximity to the edge.)

And strangely enough, in the midnineteenth century, when the development of the instruments used in physics had at last come far enough to measure one of the fixed star parallaxes and thus establish the rotation of the earth around the sun, the event generated very little interest (Fig. 1.1). A few physicists showed enthusiasm for the detailed technical and mathematical explanations that their colleagues Friedrich Wilhelm Bessel and Joseph von Fraunhofer used to confirm the beautiful new order in the heavens, but no one needed measurements to be convinced of a heliocentric world. The Copernican model had already permeated their thinking, independent of any technological advances, quite possibly by virtue of the model's harmony and symmetry. In other words, people accepted the new world-view because it was beautiful to them. They based their choice on beauty, with nothing but aesthetic criteria in mind.

Three in One

Helping the general acceptance of the new view on its way was Johannes Kepler, who as one of the pioneers in astronomy at the beginning of the seventeenth century (shortly before the Thirty Years' War), was convinced that the sun had a stationary position at the center of the universe.[4] However, just as Copernicus had done, Kepler explained his set outlook more in terms of aesthetics than science. For him, the sun embodied the "principle of all the beauty of the world,"[5] and it therefore belonged at the center. Kepler's religious needs motivated him to maintain the truth of the heliocentric view of life over and above any other view. He believed that no other vision of the world gave witness more clearly to the Christian trinity than the Copernican proposal that the sun was in the center where, while making its eternal circuit, it could give the earth light and life. In the heliocentric system, Kepler saw the incredible "image of the triune God," and as someone who viewed all scientific study as a service to God—as worship—this was what he had been seeking all along.[6]

For those of us schooled in the enlightenment, such relentless symbolism may not mean all that much. At the very least, we can understand that Kepler was illuminated by his ideas. However, the path to this illumination was much longer than we might imagine. Kepler practiced a science that did not start out with abstract concepts. Additionally, we can gather from his experiences and notes that not only external events influenced his scientific perception (observations), but also internal preparation (deep-seated images and ideas).

For Kepler, the heliocentric order of the cosmos was true (and at the same time beautiful) because it corresponded with his inner ideas, and because the images that originated through his perceptions of the outer world were in concordance with the images that, in his own words, "were lured out of intellectual and inner realities."[7] For Kepler, perception truly began when these images "light up in the soul" as he wrote in 1604. It's clear to me that this experience was a brush with beauty.

What Kepler has said about his own experience will immediately remind the philosophically well-versed reader of a thought central to Plato: Perception occurs when external impressions and inner images converge. Plato spoke in this respect of *ideas*, and how it was important to him that ideas, in contrast to fleeting and mutable appearances, be steadfast and unchanging.

Modern science tends to keep a distance from this kind of philosophical truism. In other respects modern science has been pointing in a comparable direction, most extensively in cognitive theory. Apparently, human comprehension begins long before the stage where we formulate the rational sentences we use to communicate, and it's not difficult to picture these rudimentary images. Unconscious processes operate at the outset without any recognizable concept and offer us something that can best be described as a collection of images. The mind grasps these figments of the imagination, the initial images, and compares them with the external images conveyed by perception.

The Role of Images

Each time they appear in a new work, the individual components in a painting reemerge as a completely new image; in the same way, each time perceptions successfully make their way in from the outside through the senses, they reemerge when we make images out of them. A cognitive scientist might explain this meeting of two comparable elements as the complex activity of nerve cells. He might even depict, in his technical jargon, the combining (comparison) of both image-patterns as a type of resonance that results in mental stimulation. This phenomenon was known to Kepler and other great scientists as a "flash" (which ordinary mortals experience when something suddenly becomes clear). When we need an example of this kind of inner light from the twentieth century, we can turn to Richard Feynman, the theoretical physicist. Feynman devised a graphic method called the Feynman diagram for visually understanding the interaction between elementary particles. One day when he had the diagram that correctly grasped the transformation of electrons, protons, and neutrinos before his eyes, in his own words, "the goddamn thing was gleaming."

I suggest taking such stories seriously and giving them their proper place in the annals of cognitive theory. Feynman is not being sensational when he uses this idea of "gleaming." [8] He was signaled at that moment that his diagram did in fact describe how nature functioned and that he had literally seen through it.

The Romantic poet Novalis placed the human soul where the inner and outer meet. Feynman, then, finally saw the "light" when his realization connected emotionally. Every scientist is familiar with this connection. It's even likely that there are insights that would not have been possible without feelings. When in this early state of being deeply moved, a person's feelings may best be able to discern whether or not the images there agree. I am referring to the cognitive level at which the images show something that can be guessed at but not yet named. This something sits at the core of the soul and exists without concept. Instead, psychologists refer to this something in terms of symbolic images and with these images attempt to explain how the aesthetic moment figures into the stage where perception begins.

Archetypes

One has to imagine Kepler's strange "realities," the wellspring of inner images, as essential to the human comprehension of reality and as what contributes to breakthroughs in thought. In modern psychology these images are often referred to as *archetypes*, and it is helpful to think of them as bridges between sensory perception (external) and ideas (internal).

Kepler likewise used the difficult concept of archetypes and used it in reference to our overriding topic of interest, the concept of beauty. Kepler, the experienced scientist whom science has to thank for so much knowledge, observed that our thinking begins with archetypes and that these appear visually as the geometric shapes circle, line, and angle. He wrote, "Traces of geometry are expressed in the world, as if geometry were the archetype of the cosmos."[9]

Thus, Kepler found the world beautiful because it could be understood by using geometry, as was the case with the heliocentric system. Or as generally expressed by Kepler, "Geometry is the archetype for the beauty of the world," and in the original Latin, "Geometria est archetypus pulchritudinis mundi."

As beautiful as this sentence is and as much as it deserves to be learned by heart, it is probably no longer possible for people in our century to form an accurate impression of what Kepler meant when he spoke of archetypes. In Kepler's time, the mental and the material had not yet been strictly separated, a separation that René Descartes introduced and that pervades Western thought. Ever since René Descartes, we have had to decide whether something is part of *res cogitans* or is a *res extensa*: My PC is material, my thoughts while I write on my PC are not; there is no way of expressing this neutrally. I am only able to discuss what belongs to one or to the other sphere, and there are well-grounded reasons to suspect that archetypes resist this kind of classification. For Kepler, this division was not an issue. However, the dichotomy can present a significant obstacle to anyone trying to arrive at a closer definition of archetypes.

Harmony and Symmetry

Kepler's central work is called *Harmonices Mundi* and describes "the harmony of the world." Because we can translate the title easily, we can also easily overlook its inherently aesthetic element. Kepler investigated the world because beauty drove him to it. He felt that he encountered beauty in his research, and he is not alone. Even more scientists pursued their desire to find beauty in the centuries following Kepler, up to the present day.

Great science is preoccupied with harmony, as was Kepler, as much as it is with symmetry, as was Copernicus. Albert Einstein, the best known figure of the twentieth century, brought this point home in his famous publication of 1905 in which he presented what is now known as the theory of relativity. Einstein advanced his theory because he was disturbed by an asymmetry—a lack of symmetry—that appeared to be lurking in one of the foundations of physics if someone were to put it into motion. I am referring to descriptions that showed in mathematical terms what electric and magnetic fields looked like and how they came to be mutually influential and dependent. All of this could be successfully summarized in the nineteenth century in a set of formulas known to physicists as the Maxwell equations (Fig. 1.2).

The formulas mentioned are those of James Clerk Maxwell, the Scottish physicist who, using its symbols, gave a mathematically aesthetic form to his insights into the interaction between electricity and magnetism. His British colleague Michael Faraday had arrived at the same insights earlier in a more intuitive fashion. Faraday's search for the combination of electric and magnetic forces or phenomena was likewise triggered by the desire to find symmetry in nature, that is, through an aesthetic moment. But what made Faraday begin his search for beauty?

At the beginning of the nineteenth century in Denmark, Hans Christian Oerstedt observed that a magnetic needle was influenced by a wire when a current ran through the wire (Fig. 1.3). In other words, electricity had magnetic effects. Anyone could see this, but only one person saw further, and that was Faraday. He suspected that only one side of a reciprocal dependency was visible with the deflected compass needle. Nature acting symmetrically, according to his thought, must have allowed the effect to run in the other direction as well. In other words, magnetic fields must have also been able to incorporate electric currents and potentially produce them.

Understandably, Faraday got right down to work, but it ended up costing him many years of painstaking experimentation until he could prove the symmetrical effect in the external world of wires, spools, and currents of which he had long ago been certain. The length and intensity of Faraday's search can best be understood as an expression of his longing for beauty and the conviction that he would run across it by studying nature.

The desire for symmetry also appears at the outset of the success story played out in Einstein's life in science. His famous treatise entitled *On the Electrodynamics of Moving Bodies*, which introduces the theory of special relativity, begins with the remark that the application of Maxwell's equations "leads to an asymmetry which does not appear to be inherent to the phenomena." [10]

A. Die Schreibweise, die in heutigen Physik-Lehrbüchern für Anfänger üblich ist:

$$\nabla \cdot E = 4\pi\rho \tag{1}$$

$$\nabla \cdot B = 0 \tag{2}$$

$$\nabla \times E + \frac{1}{c}\dot{B} = 0 \tag{3}$$

$$\nabla \times B - \frac{1}{c}\dot{E} = \frac{4\pi}{c}j \tag{4}$$

$$\frac{\partial Ex}{\partial x} + \frac{\partial Ey}{\partial y} + \frac{\partial Ez}{\partial z} = 4\pi\rho \tag{1}$$

$$\frac{\partial Bx}{\partial x} + \frac{\partial By}{\partial y} + \frac{\partial Bz}{\partial z} = 0 \tag{2}$$

B. Die weniger stark komprimierte Schreibweise, die Maxwell zu Beginn seiner Forschungen verwendete:

$$\left.\begin{array}{l} \dfrac{\partial Ey}{\partial x} - \dfrac{\partial Ex}{\partial y} + \dfrac{1}{c}\dot{B}_z = 0 \\[2mm] \dfrac{\partial Ez}{\partial y} - \dfrac{\partial Ey}{\partial z} + \dfrac{1}{c}\dot{B}_x = 0 \\[2mm] \dfrac{\partial Ex}{\partial z} - \dfrac{\partial Ez}{\partial x} + \dfrac{1}{c}\dot{B}_y = 0 \end{array}\right\} \tag{3}$$

$$\left.\begin{array}{l} \dfrac{\partial By}{\partial x} - \dfrac{\partial Bx}{\partial y} - \dfrac{1}{c}\dot{E}_z = \dfrac{4\pi}{c}j_z \\[2mm] \dfrac{\partial Bz}{\partial y} - \dfrac{\partial By}{\partial z} - \dfrac{1}{c}\dot{E}_x = \dfrac{4\pi}{c}j_x \\[2mm] \dfrac{\partial Bx}{\partial z} - \dfrac{\partial Bz}{\partial x} - \dfrac{1}{c}\dot{E}_y = \dfrac{4\pi}{c}j_y \end{array}\right\} \tag{4}$$

C. Die stärker komprimierte relativistische Schreibweise:

$$\partial_\nu F^{\mu\nu} = \frac{4\pi}{c} j^\mu \qquad \text{1 und 4}$$

$$\varepsilon^{\mu\nu\kappa\lambda}\partial_\nu F_{\kappa\lambda} = 0 \qquad \text{2 und 3}$$

Figure 1.2.   The legendary four Maxwell equations which express how electric (E) and magnetic fields (B) arise and mutually induce each other. Electric fields arise from electric charges, and magnetic fields through currents (j). (I'm going to skip the mathematical details for the moment.) In the vacuum in which there is no charge and no current, the Maxwell equations convert to a wave equation. It describes the propagation of an electromagnetic wave at the rate of speed c. According to the Maxwell equations this is related to the speed of light being constant. In any case, a coordinate system is not yet provided for by the equations. Einstein took this property so seriously that he altered Newtonian mechanics to make the speed of light constant. We call the results the theory of relativity. A. The notational style commonly found in current elementary physics textbooks; B. The less compressed style used by Maxwell at the outset of his research; C. The extremely compressed relativistic style.

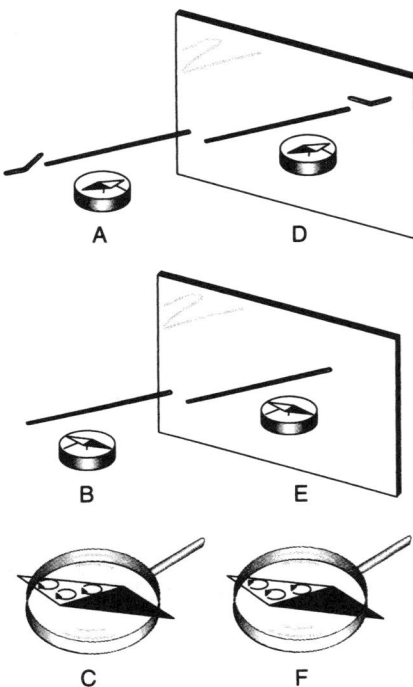

Figure 1.3.   Oerstedt's experiment looks very simple but has a few problems with symmetry. In the first place the compass needle (A, D) turns to the left when the current flows (B). We know today that the interaction is produced by the magnetic field built up by the current according to the Maxwell equations. (See Fig. 1.2.) When we observe the procedure in the mirror, something doesn't seem quite right: At first glance, the deflection of the needle takes place in the "wrong" direction (E), which raises the question of whether the mirror image of the experiment is showing something that is not physically possible. The answer is that the reflection not only switches left and right but also north and south poles because the magnetic properties of the needle are produced by rotating electrons, and their sense of rotation turns them around backward (C, F). Symmetries are not as obvious as they seem. In order to perceive them, one has to have already started thinking about them.

Whereas Faraday was faced with the contradictory effects of electric and magnetic fields, Einstein was dealing with a conflict between the electromagnetic Maxwell equations and the Newtonian laws of mechanics. In the mechanics paradigm, Newton's speed of moving bodies added up without a hitch; in contrast, the Maxwell formulas allowed a deviation. As a result, the speed of electromagnetic waves composing light had to remain constant. This meant that the speed had to be independent of whether or not the source of the rays moved. The light of a flashlight would keep moving at the speed of light no matter how fast the operator holding the device ran from place to place. To shed light on the world and see it in the future, you would have to be riding a wave of light.

The Maxwell equations must have made a big impression on Einstein. Despite the fact that its conclusion—the constant speed of light—was far from obvious at first glance, flatly depicted, and irreconcilable according to the best intuition, Einstein elevated it to the principle that turned the world of mechanics upside down and fundamentally changed our understanding of time and space. Einstein saw the asymmetry between Newton's mechanics and Maxwell's electrodynamics and he decided to hold on to the equations that were more beautiful and to throw out the laws governing motion—by changing them. (In autobiographical notes, Einstein later apologized for this to Newton.)

In the Maxwell equations—and this is a part of what makes them wonderful and what enraptured physicists—the speed of light occurs twice: once as the speed with which waves of light are propagated in empty space and again as a fixed numerical value that is necessary to convey differing units of measurement. Einstein drew the conclusion that the speed of light must in fact be constant, and though without reliable experimental evidence, he trusted in the power of his conviction. For what seems to have been purely aesthetic reasons, he then spent many years transforming and reforming the remaining laws of physics—Newtonian mechanics, for example—so they could be reconciled with the unchangeable speed of light.

Aesthetic reasons more than any others led Einstein to the theory of relativity, a motivation that should begin to make clear to the lay person how physicists could be as fascinated with the Maxwell equations as musicians could be with a full musical score. The magic that emanates from this quartet-like formula is best conveyed by Viennese physicist Ludwig Boltzmann. Boltzmann described how once when he was looking at the Maxwell equations he felt as much joy and delight in life as Goethe permitted his Faust when he opens the book of Nostradamus.[11] In the monologue at the beginning of the universal drama, Faust sees the "sign of the macrocosm," as it states in one set of stage directions, and after one look, he cries out:

> Was it a god who wrote these signs
> that still my inner storms,
> fill my poor heart with joy,
> and with whose mysterious impulse
> the powers of nature are unveiled around me?*

What Boltzman perceived in the nineteenth century while viewing Maxwell's theory of electromagnetic fields, physicists in the twentieth century perceived anew when Einstein revealed the theory of relativity concerning the space–time continuum. Aesthetics had not only motivated his search, it also led him to a beautiful finish.

"This Special Beauty"

We can begin to understand how great science operates if we accept that its creators are going through an experience very similar to that of poets and painters. Henri Poincaré, the French mathematician, expressed this most clearly in response to the newly published theory of relativity, when he stated in his book *Science and Method*, "The scientist does not study nature because it's useful: He studies it because he takes pleasure in it, and he takes pleasure in it because it is beautiful. If nature were not beautiful, it would not be worth knowing." [12]

There is almost nothing to be added to this until a few lines later when we run across the comment that Poincaré is not "speaking of that beauty which strikes the senses," and thus not "of the beauty of qualities and appearances." He means the "more intimate" beauty that "proceeds from the harmonic classification of parts and which can be grasped by pure intelligence." Poincaré now prescribes accordingly a search for "this special beauty," and as the French mathematician and philosopher observes, it is this search alone for which the "scientist condemns himself to painful labours." [13]

---

*War es ein Gott, der diese Zeichen schrieb,/Die mir das inn're Toben stillen,/Das arme Herz mit Freude füllen,/Und mit geheimnisvollem Trieb/Die Kräfte der Natur rings um mich her enthüllen?

We can agree to this last sentence without reservation, but if our goal is to trace aesthetic moments in science, we won't get very far without the kind of beauty that strikes the senses. When Poincaré said aptly that we study nature because its beauty brings us joy, he was departing only slightly from the first lines of that famous book noted for the beginning of Western science, Aristotle's *Metaphysics*, which begins with the declaration most often quoted in philosophical address, "All men by nature desire to know."[14]

Aristotle, who of course was only referring to people within his own cultural sphere, explained his statement by the second sentence of his *Metaphysics*, one rarely mentioned any longer in most philosophical addresses. According to Aristotle, we strive for knowledge not out of a lust for power, but out of joy. Aristotle even states the source of joy: "All men desire to know. An indication of this is the delight we take in our senses; for even apart from their usefulness they are loved for themselves; and above all others the sense of sight."[15]

On Perception

The human search for knowledge starts with perception, and this ability, rather than the pure intelligence of Poincaré, helps us see a thing, for example nature, as beautiful. The ability to perceive is the first step on the way to real thinking and is one of the criteria used to single out a great scientist from the crowd. When Faraday saw the compass needle turned by a current, he was not only observing the needle's turning; everyone could see that. He was perceiving much more—in this case that electric and magnetic fields (remaining invisible to the eyes) were working in harmony, an occurrence he was to investigate further. Similarly, when Einstein saw a falling object and imagined himself in its place, he wasn't just noting the height and time span of the fall, something anyone could do, but it had suddenly become clear to him—he perceived it—that a person in the midst of a free fall doesn't feel gravity anymore, a phenomenon for which he then sought the underlying reasons. After many years of searching, he discovered the general theory of relativity.

When Aristotle speaks of perception in his *Metaphysics*, he uses the Greek word *aisthesis*. We can easily see that our present-day concept of aesthetics is derived from this word. As apparent as the association may be, it is just as apparent that the old and the new usage of the word *aisthesis* have little in common. Today when we talk about aesthetics, we aren't necessarily thinking of perception and cognition but of a theory of beauty in reference to works of art. The task of this modern aesthetics is to answer the question, "What makes a work of art *art*?" or "What is the difference between a wine rack and a surrealist sculpture?" Artistic aesthetics were expressed concretely for sculptor Joseph Beuys, known for his use of unorthodox materials, when a chunk of fat he fashioned underwent significant alterations in appearance after the cleaning personnel made their rounds.

I do not mean to interfere with aesthetics in the arts. By the same token, we should not abandon the attempt to reclaim the beauty of aesthetics for the natural sciences, the beauty we can grasp intellectually using our operative reason and understand sensorially using our perceptive ability. Not only works of art but also scientific treatises in mathematical language can be said to be of "unusual beauty" as Ludwig Boltzmann once expressed it in an unapologetically lofty speech around the end of the nineteenth century:

> Beauty, I hear you ask; doesn't its grace flee away as soon as integral calculus strains its neck, can something be beautiful when the author lacks the time for even the smallest embellishment? But no, it is precisely the simplicity, the absolute necessity of each word, of each letter, of each little stroke which of all the artists puts the mathematician next to the creator of the world; the mathematician establishes a grandeur that has no equal in art, that has at most a similarity to symphonic music. The Pythagoreans already recognized the similarity of the most subjective and the most objective of arts—Ultima se tangunt. And what a capacity for expression, what subtle characteristics mathematics has. Just as musicians know their Mozart, Beethoven, and Schubert, within the first few beats, so the mathematicians know their Cauchy, Gauss, Jacobi, and Helmholtz after a few pages. The greatest degree of elegance and occasionally flimsy deductions characterize the French, while the English are known for their dramatic powers, especially Maxwell. Who isn't familiar with his dynamic theory of gases? At first the variations of speed develop majestically, then from one side the equation of state is established and from the other side the equation's central movement; the chaos of the formula sways ever higher; suddenly the four words resound: "Put $n = 5$." The evil demon V disappears, and as in music a wild figure of the basses undermining everything up to now suddenly goes silent; as with a stroke of a magic wand what moments before seemed indomitable all comes to order. There is no time to say why this or that substitution was made; anyone who can't feel it should put the book away; Maxwell is no incidental musician who has to put explanations over the notes. Pliantly

now the formulas spit out one result after the next until the surprising finale—thermal equilibrium of a heavy gas is achieved and the curtain falls.[16]

A Question of Form

We don't need to understand all the particulars to realize that what Boltzmann was trying to express here is representative of all heartfelt scientists. However, it should be clear by now that at least for both the savoring critic and the aesthete, scientific work is distinguished not only by its content, but just as much by its form and style. It is always the style of presentation, the form of a work, that reflects its author. Scientific publications tend to have the most penetrating influence on research when in addition to significant content, they have a striking form. Of course in order to be accepted, any work delivered for publication in the realm of the exact sciences, especially in the current century, must observe predetermined specifications. There must be a summary; the method must first be designated and the results produced before the significance of the entire text can be debated in discussion. Obviously all of these formalities harm form by not allowing it very much flexibility. In spite of this, great authors, in the sense described by Boltzmann, manage to use their style to excel and to adapt the form of presentation to lend their conclusions force and persuasion. It is this attention to aesthetics that elevates them far above average and permits them entrance into the heavens of the classics.

An example is the famous work by the American James Watson and the Englishman Francis Crick that they submitted in 1953 to the scientific public as a proposal for the structure of genetic material in the trade journal, *Nature*[17] (Fig. 1.4). In 1953, Watson and Crick first presented the double helix, which today textbooks and illustrations in every imaginable magazine have made familiar to the average child. The double helix has become the symbol of the age of molecular biology, the age it has itself ushered in.

There are many scientific reasons to consider that short work of 1953 a great turning point in biology, and anyone who picked up the original text would be hard-pressed to overlook the aesthetic aspects. Of course, after 1953 there were still many other proposals for the structure of genetic material, and there are certain variations that Watson and Crick did not consider or understand. In spite of this, their work spread throughout the

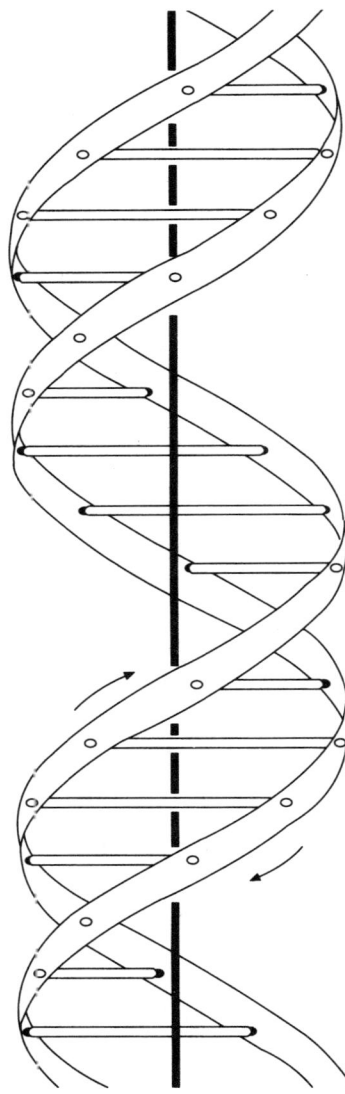

Figure 1.4. The original double helix from the journal *Nature* in 1953 was originally drawn by Odile Crick. She was interested in its structure not as a scientist but
as an artist. I imagine that the aesthetic quality of the molecule accounted for the
publication's lasting impression. The structure must not only be right, it must also
be presented in the right form.

scientific world with an unusual air of finality, and this was due not only to the words they used in their presentation, but also to the more persuasive than precise form of the double helix itself; the new model of a genetic molecule was the work of a graphic artist. The first double helix was drawn by Crick's wife Odile who had already made a name for herself as a painter. When we see the structure of genetic material through the eyes of a painter, our senses, that is, our perception, responds. The beautiful spiral is immediately satisfying, and no change or correction to any of its details will influence its power. These minor details hold no interest simply because they are not beautiful.

The Beauty of the Spiral

It is impossible not to be fascinated by the double helix, and although it is probably hasty to think we can fully classify every aesthetic aspect of this miraculous achievement of evolutionary nature in an appropriate form, I am going to attempt just this (Table 1.1) without going into any further detail about the reasons for the molecular spiral. Even the losers in the race for the illumination of structure, Watson and Crick's competitors, noticed the aesthetic qualities of the DNA spiral. Even they weren't able to escape the beauty of the model and were overheard saying things like "too beautiful to be false."

Not only the view of the elegant exterior (Fig. 1.4) of the molecule from which life originates and on which it depends is beautiful. The view into

Table 1.1.   Some Reasons
for the Beauty of the Double Helix

Reason	Details
Artistic sketch	p. 16
Ratio of the golden section	p. 54
Pythagorean tetraktis	p. 33
High degree of symmetry	p. 116
Circular form of the cross section	p. 18
S-lines	p. 38
Unity in variety	p. 44
Experimental elegance	p. 22

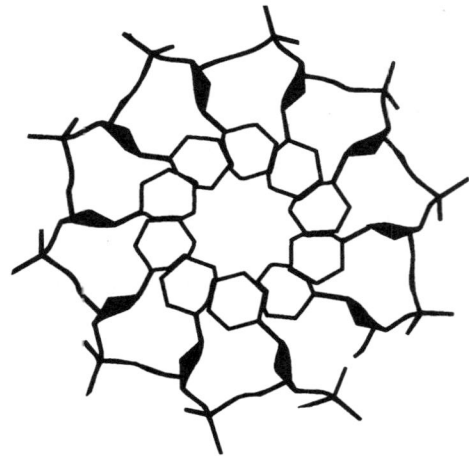

Figure 1.5.   The interior view of the double helix. The view of a DNA molecule along the axis.

the structure's interior as it winds into itself (Fig. 1.5), where the structural components, shaded just right, can be made to resemble the rosette of a gothic cathedral, is also beautiful.

Further, we find the golden section within the genetic spiral, and according to the chemical properties of the genetic molecule, the structure's interior has exactly four components. It would have pleased Pythagoras and his followers, the Pythagoreans, that the holy quaternion that they inherited and named *tetraktis* would be bound right to the core of life and would spread its influence from there.

These comments and lists of reasons are not meant to falsely attribute magic or other unscientific characteristics to the double helix. They are meant only to express that it is wrong to say that due to our knowledge of this structure, there are no more secrets in the land of the living. In fact the double helix does nothing to solve the puzzle of life—the double helix is the puzzle of life. It's only a matter of perceiving it in the right way.

The Relevance of the Beautiful

The English physicist Paul Dirac once said that Einstein's theories convince us not only by their truth, but above all by their beauty.[18] If observed astronomic data proved that Einstein's theories do not grasp the physically measurable reality of the cosmos accurately, according to Dirac, it wouldn't disturb a single person's current view. In general, Dirac found that a false but beautiful theory holds more interest than an ugly but accurate one. The conclusion he drew personally was to no longer worry about a modern form of physics, or quantum electrodynamics, as much as to be able to predict the results of experiments. According to him, this theory of the interaction of light and matter lacked every kind of elegance and beauty.

What Dirac said about Einstein and others transfers without difficulty to Watson and Crick's model. As a painter, Odile Crick created an image that convinces us immediately that it is an accurate representation of how the genetic substance is built, and no one would want to believe otherwise. Detail-obsessed chemists who threaten the model with small deviations to make it match reality may have to reckon with half-filled lecture halls, and they must be prepared to be dismissed in a less than scientific way. After the double helix was named, for example, Max Delbrück, Nobel prize winner and my doctoral advisor, counseled all those present at the very beginning of a lecture in which an alternative structure was to be offered to go home and spend their time in a more sensible way. In the end the double helix was not to be doubted.

At least the special form of Watson and Crick's 1953 work takes the wind from the sails of the argument often used to place the scientist on a lower rung of importance in comparison to the artist or that supports the view of scientists as interchangeable while the unique genius of the artist is irreplaceable.

It is commonplace to hear that the double helix would have been found even if there had been no Watson and Crick. Some scientist or other would have been bound to think of arranging the molecules in a staircase or a spiral. In general, we think we can say in terms of science: What Doctors A and B have discovered today, Professor C will discover tomorrow. On the other hand, so goes the argument: Goethe's poetry or his Faustian tragedy would never have existed without the poet.

Such arguments are raised out of different motives, but for the most part are intended to disparage science and its activities. And yet, this is the very reason the comparison limps along. In the case of the poet, it's a reference to the form given the material. There were poems about love and fidelity before Goethe, and other dramatists have worked on Faust material before and after. Whereas in the case of the scientist, the content is addressed, all of which applies to the structure of the molecule. If we compare what is actually comparable—the form—then the form in which a scientific publication presents its results might not be treated derogatorily and dismissed as incidental. The form stands as an achievement in itself and has a thorough influence on the effectiveness of the results.

This brings us back to Dirac's assessment of a theory's beauty that explains why the public and many scientists cling for so long to theories even when they know they are false. For example, if we look at it from a strictly scientific view, the heliocentric point of view bound up with the great names Copernicus and Kepler is false. We know that we cannot say that the sun is "resting" at the center of the universe. And just as false is the notion, always a favorite, that electrons move on well-defined orbits surrounding an atomic nucleus that have a fixed position and structure, as suggested by the famous *Atomium* in Brussels or various other drawings. People are drawn to these models because they are aesthetically appealing; I see no other reason. Science is accepted, apart from all requisite correctness, when it generates a little beauty; on the other hand, when it fails to do this and turns out ugly, it is dismissed. Interest in science, it seems, depends not primarily on the truth embodied by its results, but instead on the aesthetic qualities it provides the public.

The Elegant Experiment

Form becomes a critical element in science not only in presentation style. Its influence is also apparent in the results of both beautiful theories and elegant experiments. Elegance is actually the highest praise that can be given among scientists, and despite what this aesthetic classification would suggest, they have in mind the knowledge gained by the particular experiment. Its value is based on its elegance.

For example, let's look at the explanation for a question posed in connection with the discovery of the double helix. Genetic material must have the property of duplication, and the double helix clearly supports the hypothesis that from one double strand of the molecule two are made. But how? Do we find in the second generation an old and a new double strand? Or would there be two double strands, one half old and the other half new? But above all: How can we design an experiment to answer or make any decisions in terms of this question?

The solution achieved by two American biologists, Frank Stahl and Matthew Meselson, and their 1957 Meselson–Stahl experiment was, in trade circles, an immediate classic, in plain English: it lacked nothing.[19] Its form was perfect and therefore also perfectly beautiful.

The year-long antecedents, the search for a suitable system for the planned experiment, the actual activity in the laboratory, operating with analytical ultracentrifuges and radioactive chemicals and much more is about as elegant as a painter nailing together a canvas frame or cleaning brushes. However, the product of all these efforts resulted in such a lucid and corresponding form (Fig. 1.6) that many scientists today still sound utterly fanatical during their lectures on the famous Meselson–Stahl experiment.

One condition of this experiment consisted in making the genetic material physically heavier without changing it chemically. There is something beautiful in this idea alone, the understanding that the chemical properties of an atom—for example, its ability to bond with other atoms— are determined by its external electrons, whereas the physical properties— for example, the mass—are hidden inside the atomic nucleus. This distribution became apparent through the presence of atoms in which the nucleus was a little more voluminous, with the electrons following suit.

Experts talk in these terms of the isotope of an atom, and Meselson and Stahl began in October 1957 by feeding bacteria nutrients containing a heavy nitrogen isotope. This nitrogen was gradually incorporated into the bacterial molecules through metabolic processes, and both biologists continued this treatment for as long as it took for the entire genetic material of the bacteria to become heavy. They then began to feed it normal nutrients (e.g., sugar) with the naturally light nitrogen, and examined the bacteria of each subsequent generation's growing genetic material at intervals.

In fact—and this was a further requirement for the elegant experiment— for the first time a highly subtle method stood at the disposal of the scientists—analytic ultracentrifugation. Thanks to this new technology they were in the position not only to differentiate heavy and light genetic material, but also the in-between stages. The image that the interpretations of this procedure revealed, and the image that one could make following the actual process itself (Fig. 1.7) are so closely related to each other that their aesthetic charm and their power to persuade, perceptible through the senses, allowed the Meselson–Stahl experiments to speak for themselves and made all further commentary superfluous.

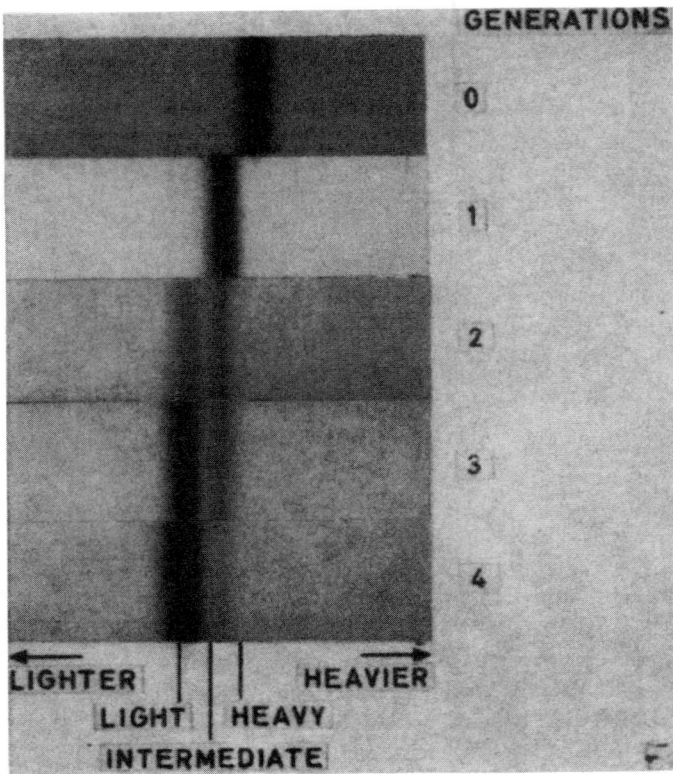

Figure 1.6.   The result of the Meselson–Stahl experiments from 1957 consists of dark bands that change from one generation to the next. Over time the genetic material becomes lighter by degrees, as shown by the shift toward the left. The experiment begins (Generation 0) with cells (of bacteria) grown in a medium with heavy nitrogen (details in text). The following generations are developed with ordinary (light) nitrogen. The corresponding mechanics, as conceived, are shown in Fig. 1.7 from John Kendrew (1966), The Thread of Life; an introduction to molecular biology. Cambridge: Harvard University Press.

A Beautiful Swindle?

Further back in our discussion I mentioned Paul Dirac's dictum that a beautiful theory is more interesting than a correct one. I am certainly not saying that fully false, and therefore absurd, theories as elegant as they may be can exist under the nose of science. However, in spite of this, most

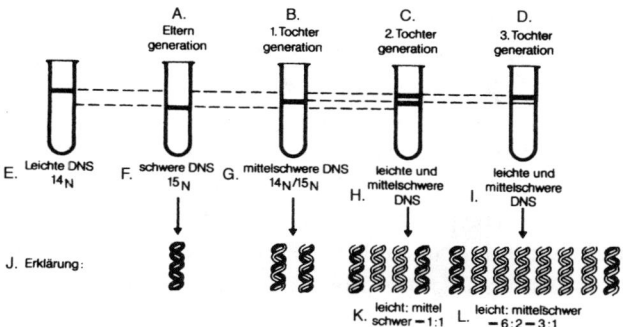

Figure 1.7. A presentation of the processes that become visible through the Meselson–Stahl experiments (Fig. 1.6). We can see the accepted course of replication, directly readable from the experimental result. For most scientists and the public this is not the end of the story. We also have a beautiful image of the fundamental mechanism. A. parent generation; B. first daughter generation; C. second daughter generation; D. third daughter generation; E. light DNA ^{14}N; F. heavy DNA ^{15}N; G. intermediate DNA $^{14}N/^{15}N$; H. light and intermediate DNA; I. light and intermediate DNA; J. explanation; K. light: intermediate = 1:1; L. light: intermediate = 6:2 = 3:1.

scientists do regret when their beautiful thoughts and theories are refuted by ugly facts. It's sometimes hard to resist the impression that they would rather see past the facts than to come up with an idea that might compel them to change their notions and theories.

If we were to test the philosophical capacity of this last sentence, we would quickly be forced to investigate the nature of the fact. *Act* takes up the better part of the word, and in experimental research a *fact* is elicited from results, that is, an *act*ion someone has performed to elicit facts. It goes without saying that an experiment can be conducted poorly or can contain mistakes, and in this case we can easily deny the fact—the result—if it doesn't suit us.

The American physicist Robert Millikan operated according to this motto when he tried to determine the charge of an electron in the first decade of the twentieth century. One speaks of the elementary charge and means that electric charges in nature are not distributed continuously in every magnitude, but instead only as whole number multiples of a least value—the elementary charge itself. Today this understanding (with limitations that have no bearing here) is indeed treated as a fact, but when Millikan set to work on his famous oil drop experiment to measure the charge of an electron, this couldn't yet be said. Millikan could only accept that he was on the heels of the elementary charge.

The way in which he proceeded is that much more surprising. The lab notebooks that only became well known after his death show that Millikan only published experiments that corresponded to his theory of elementary charges.[20] When they did agree, he even used the word *beauty* to characterize the corresponding measurement (Fig. 1.8). The Nobel laureate only published the beautiful data—he shut his eyes to the rest!

Did he beautify his data or hold it back? Was beauty used to betray? This is one possible opinion and as such deems Millikan's actions as simply a piece of scientific trivia. I suggest taking Millikan's *beauty* and his enjoyment of it more seriously. It seems obvious to me that a successful and influential physicist was depending more on his feeling for the beauty of nature and less on his analytic and technical ability—and indeed exactly twice. The first time, the thought of an elementary charge was so beautiful to Millikan that he undertook an experiment on its account. The second time, we have to trust him as an intelligent and experienced physicist who was able to discern in the course of a measurement whether or not it was useful.

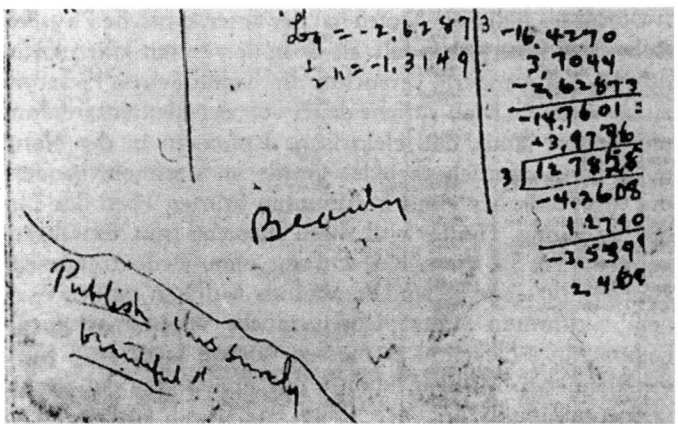

Figure 1.8.   Excerpts from the lab notebooks of Robert Millikan that show in their entirety that the American physicist didn't keep all the data but instead only searched out and published the small part that was beautiful to him. The word *beauty* can't be overlooked. What was "beautiful" was published.

The actual question is found at another point, which concerns where Millikan found the inner confidence and conviction that an elementary charge existed at all. He could only act the way he did because to him the existence of such a constant in nature was obvious in spite of all rational doubt. This is only understandable when we characterize these thoughts as beautiful and assume that scientists investigate nature because it is *beautiful*. This is precisely what Millikan did in his own way—not more and not less. He found the beauty he longed for and stated it expressly.

2   *All that Beauty Can Be: The Lengthy Consideration of a Simple Concept*

Beauty is . . . I can not say
suffice its forms are many [1]
Albrecht Dürer

In the first chapter, when I made the point that throughout the entire history of the sciences in the centuries from Nicolaus Copernicus to Henri Poincaré scientists have tried not only to discover the right things but also the beautiful things, I was raising the obvious issue of whether the central concept of this book, which I am turning around like a planet around the sun, can be classified, defined, or evaluated in any way.

What is beauty? And when do we use the slightly different concept of *beautiful*? How does beauty in nature, say the contours of a landscape or the appearance of a flower, differ from the artificial beauty that human beings create? And what about human beings themselves? What lends them or their form beauty? Can we define what a beautiful body is, and if we can, can we do this once and for all? And do we understand why we can talk about beauty in nature and the beauty of a view of life or a theory without danger of slipping into absurdities?

Most of these questions must remain unanswered because there are no and will never be databases or Internet addresses where the answers we are looking for are saved for easy reference. It's useful, however, to consider the ways in which we use the word *beautiful*. We use the word *beautiful* to name objects found both in nature and in art, and we distinguish them with this concept when they appeal to us. We also use the word *beautiful* to name the sensation the appealing object gives us that wakes a certain desire or sense of pleasure. Maybe it's easiest if we imagine beauty as the special quality of perception that releases in the observer a feeling of well-being and happiness and that compels him or her to find that which is worth seeing as worthy of love.

The Given Beauty

In any book other than a philosophy book or an art history book, statements may be made about the beautiful without distinguishing it from the other form of the word—*beauty*. The beautiful becomes beautiful through beauty, a famous philosopher was to have once said—a beautiful sentence. I will use the two concepts, however, with as close a meaning as possible because the distinction between the two is not particularly relevant to scientific discourse.

It becomes clear that there are different approaches to this subject when we consider Poincaré's intellectually comprehensible beauty and the familiar beauty that we perceive through the senses. I would like to suggest that no one should give up their own impression of what is beautiful or what the beautiful might be. Besides the use of slang expressions (what a beaut!), when we hear people use the attribute beautiful, it's clear that a grasp of the concept requires little training. As adults, we should never allow anyone else to determine what is beautiful to us, if only because no court or persons of higher authority exist that can do so: Beauty is what pleases apart from any concept, as Immanuel Kant expressed it, who was not only familiar with the taste of his enlightened contemporaries, but who at the same time reminded philosophers that those who want to experience beauty should feel free to do so without needing secondary motives.[2]

Kant's statement seems to suggest that beauty is not conducive to comprehensive explanations or statements in terms of other relative qualities of objects, such as the glitter of gold or of diamonds, which then become synonyms for beauty. Attempts of this kind abound, but no one ever notices that the new concept doesn't get any easier to grasp (and everyone knows that all that glitters is not gold). When I say, for example, that beauty is that which glitters and think I've explained what beauty is, I am clearly mistaken. Now, instead, I'm confronted with a new task, to illustrate what "glitter" means and how it releases a pleasant reaction in the person perceiving it. I haven't simplified my definition; from the scientific viewpoint, I've made it more complicated.

Maybe all that's necessary is to trust that beauty or "the beautiful" exists, that everyone can recognize and sense it, and that it is simply something we have been given, that it is not our job to search it out. This may apply for truth, which was called "the unconcealed" by the Greeks, meaning only that it is generally concealed (from the senses). Beauty, on the contrary, is never concealed. The form it takes for us is plainly visible. We can perceive beauty with the senses as well as with the mind, and we acquire this ability in early childhood when we recognize our parents and comprehend their beauty aesthetically. From the very beginning of our lives we are drawn to beauty. We perceive the world from this perspective before we wrap up reality in the concepts with which we then communicate it. We are originally aesthetic beings, and it might be worth finding out how to use this capacity fully, with the joy that goes along with it.

The fact that beauty is given freely, *gratis*, and that it is a grace, *gratia*, is one of Theodor Haecker's points in his "Versuch über die Schönheit" that appeared in 1936.[3] (He was banned in the same year by the Nazis from speaking and publishing.) Haecker expressed doubt in his essay that a person is the one who lends any particular object beauty it wouldn't otherwise possess, for when we talk about landscapes as beautiful we are not just expressing our current mood. According to Haecker, even if we can't define exactly what it is about the scene that makes it beautiful, we are still making a statement about the landscape. There is nothing beautiful about a landscape. There is no detail that constitutes its beauty—either the color of the meadow or the contours of the promontories. Only the landscape itself, as a whole, that our perception takes in with the senses, independent of any concept, deserves to be assigned this coveted attribute.

Neglected Beauty

We can try for a more precise or more dependable explanation of the concept of beauty, as philosophers have attempted throughout the course of Western culture. However, as soon as we try to trace our thoughts about beauty, we find in the course of our reflection that many ideas have too carelessly drifted away from what we actually perceive, and what we perceive, more than what we ponder, is responsible for knowledge and understanding of the world.

When we look to philosophers for revelations on beauty, we find that many German philosophers started thinking more about beauty in terms of art and of the artist after they read G. W. Hegel's characterization of beauty as the sensory shining of an idea.[4] As a result, they all but exclude beauty in nature. Hegel's definition is reminiscent, if only remotely, of the true beauty Socrates offers us in the Platonic dialogues when he views earthly beauty and feels moved by it. However, as fine as this definition may be, it's probably not going to help us. Additionally, if we look at the history of philosophy starting with the Greeks, we find that many philosophers, René Descartes, George Berkeley, and Baruch Spinoza for example, treated beauty in a more or less offhand way without getting any closer at all to defining it.

In other words, there's nothing to stop scientists from approaching beauty if they're drawn toward further investigation. Further, this freedom is worth using, especially in terms of the modern science that from Europe in the seventeenth century managed to gain the confidence of the Western world and today has practically conquered the entire globe. This new era in science is distinguished by a central intellectual achievement— discovery of the object. When we practice science today, a subject observes an object, and a subject strives to report as objectively as possible and make objective statements.

Consequently, we have a difficult time avoiding the famous controversy as to whether beauty is subjective or objective, and whether the beautiful originates with the object or with the observer. Is there beauty in the world even when there is no "I"? Or does the beautiful only take place in the mind of the observer as a phenomenon conditioned and made possible by the senses?

In my eyes the answer, as Theodor Haecker suggested, can only be "both." In other words, beauty takes place between the observer and the observed. It brings and binds both together. Beauty also brings people closer together. We want beauty. We try to get closer to beauty. We set our sites on beauty even when we are investigating nature.

Essentially, beauty is a positive value for which we strive. We can say that beauty pleases without being able to say exactly why. Beauty brings pleasure, although obviously not everything enjoyable is beautiful. Beauty enables us to love and motivates us to act. Each person experiences beauty individually, through individual perception, as he or she turns it attentively toward the world and toward people in an attempt to understand them.

In Praise of the Fuzzy

Even if we use definitions of beauty with which anyone would agree, we still need to ask if we require an exact definition of beauty to be consistent with scientific goals, because in science it's necessary to clearly define each concept used. Just to engage another person in scientific discussion we have to use well-defined terms and make sharp distinctions between concepts. How else are we to escape the danger of the arbitrary?

In this case, however, I would answer that the conceptual sharpness usually called for is unnecessary. I would even go so far as to maintain that the definitions for standard concepts of science are fuzzy, and in fact must be fuzzy in order to make scientific dialogue possible at all. In my opinion, the desire to use every conceivable means to make a concept more precise is, in the scientific process, an even bigger mistake.

First, we can't make any progress when researching a concept—such as the atom, the gene, or life—when the precise meaning of the concept is established forever and for all time from the outset. Even by glancing briefly at the history of science, we see that "atom" means something completely different today than when the concept was coined in classical antiquity.

Second, there is no point in defining a concept so narrowly, the gene for example, so even an engineer is satisfied. For the engineer's purposes, it's enough to define the gene as a molecule of a certain length in the known and loved double helix shape, even though this definition isn't very useful to the evolutionary biologist or the immunologist. The latter are less interested in the naked genetic material than in its interaction with other materials in the cell and how it manages to remain stable long enough to appear as a gene at all.

Third, as soon as we closely define a concept, it suddenly becomes easy to find counterexamples, especially when better techniques for measurement arrive on the market. As a result, we end up looking ridiculous. When something like *purity* is defined as the complete absence of foreign substance, we can draw the immediate conclusion that purity does not exist in any practical sense. Some kind of analytical process is bound to ensure that a mixture will be found.

Fourth, a sharply defined concept will be remote from the sensory perception where it originated. We can set up a precisely determined system and try to gain insights from it, but to use this system surrenders sensory knowledge and as a result can never be successful.

Beauty Is . . .

After this protracted intermediary discussion, no one will be surprised that there is no exact definition of beauty in the natural sciences. Maybe we can talk about beauty in the same sense that Newton talked about gravity: What it is exactly, no one can say; it's enough for us to know that it exists and that we can explain how it acts.

In the course of Western history, however, many people—even outside of the philosophical system—have thought about how to understand especially striking or particularly effective aspects of beauty. I would like to consider a few suggestions here with the ulterior motive of seeing whether they are productive in the context of natural science or whether we are better off without them.

The definition, which for our purposes is the most elementary, has its origins in the groups of disciples who flocked around Pythagoras. Around the fifth century B.C. they searched for harmonies and discovered that numbers were a means of expression, coining sentences such as "all things are numbers" (Fig. 2.1). For the Pythagoreans, there was more to numbers than quantity. For the philosophers of this school, numbers revealed qualities. Numbers could be holy, as the number four that was called *tetraktis*. Its meaning is manifested by the four elements earth, water, fire, and air that make up the classical periodic system, or in the four

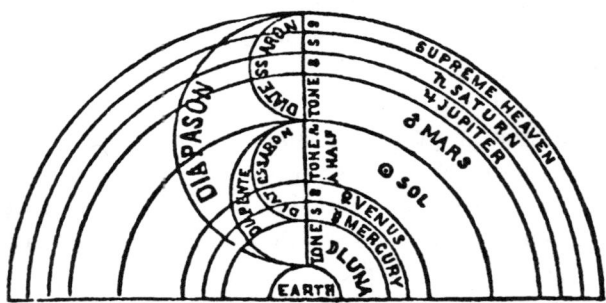

Figure 2.1.   Pythagoras is known today for the theorem for whose proof he was said to have sacrificed 200 oxen out of gratefulness to the gods. Pythagoras and his students at the time discovered among other things that they could reach harmonic tone intervals on a monochord with whole number ratios of the length of the strings. His ideas on the harmony of the spheres led him to see numbers and number ratios as the shaping form of the cosmos. He founded the Pythagorean school that swore by the holy number four. Pythagoras also presumed earth and life to be round—that they form a circle—meaning he believed in reincarnation. In any case we can find in this school of thought some archetypal ideas (circle, four) that have not lost their charm and are considered beautiful. In the diagram, Pythagoras tried to construct a relationship between the scale (in whole and half steps) and the positions of the planets found between the earth and the sphere of the fixed star.

humors of Hippocrates (blood, phlegm, and two kinds of bile) that a healthy person must keep well balanced. A Pythagorean would notice right away, for example, that the structure at the very center of life, the double helix, has a holy tetraktis at its core made up of four chemical compounds (called base pairs) that make life possible (Fig. 2.2).

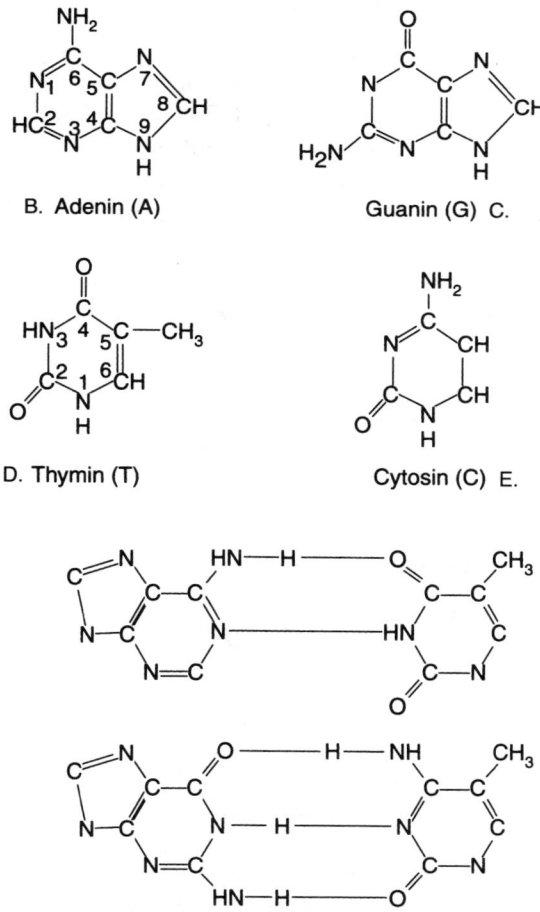

Figure 2.2. The double helix is made up of four bases that appear as two base pairs. In the molecular center of life we find the tetraktis as a fundamental component. Pythagoras would have liked this. I know I do. A. Genetic material; B. Adenine (A); C. Guanine (G); D. Thymine (T); E. Cytosine (C).

The followers of Pythagoras defined beauty primarily through external measurements: A formation was considered beautiful when the parts of the observed structure and the whole were in harmony with each other. For example, the mysterious symbol of the Pythagorean school, the pentagram (Fig. 2.3), illustrates by its sides and diagonals the number defined as the golden number or the golden section (which gets its own chapter later).

Pythagoras had an influence on Plato that is not to be underestimated. Plato was not only a philosopher, but also a geometer, and Pythagoras's influence is especially evident when it comes to Plato's explanations of the material world, as we refer to it today. He drafted his famous regular solids (polyhedrons), and there were once again four pieces, which classical antiquity since Democritus has used to construct the four elements that make up perceptible reality (Fig. 2.4). It may seem absurd at first glance—how can the universe be made entirely out of geometrical formations? It gradually begins to make sense, however, within the scope of modern physics, if we interpret Plato's effort as an inquiry not into the material foundations but into the aesthetic foundations of the world found in the form of symmetry. Only symmetrical bodies, for example, the cube, tetrahedron, octahedron, and icosahedron, were permitted in Plato's imagined

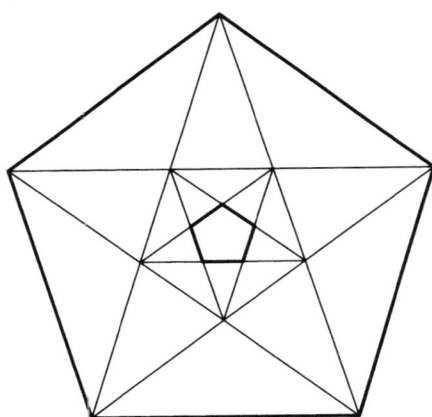

Figure 2.3.   In a pentagram, the diagonals cut across reciprocally with the ratio of the golden section, which is considered beautiful (discussed at length in the next chapter). In the interior, another pentagram is formed and the same is true for its diagonals. In principle, the process can be repeated infinitely with an endless amount of self-similar pentagrams. We can also imagine the process in reverse, that large forms grow out of their small predecessors.

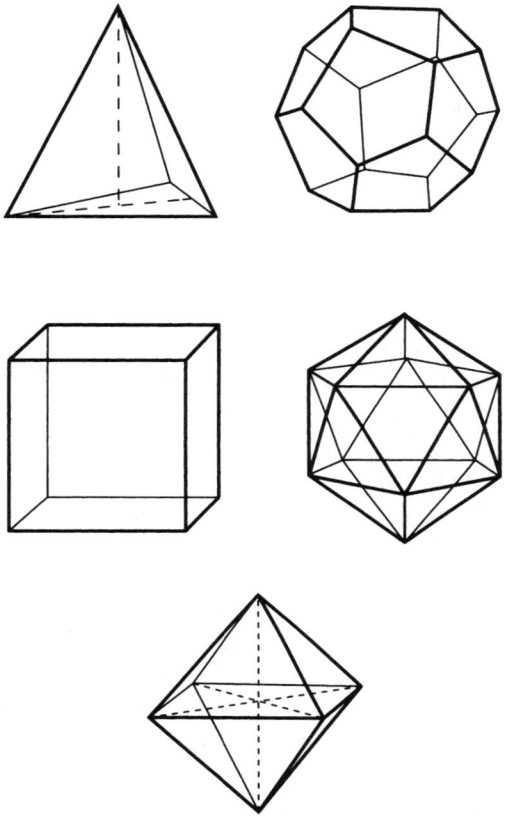

Figure 2.4.   In the dialogue "Timaeus," Plato classifies the basic components of the world as regular solids—earth, the cube; air, the octahedron; fire, the pyramid tetrahedron; and water, the icosahedron. According to Plato, the creator (demigod) designed the world in the form of a pentagonal dodecahedron. Therefore there are five universal bodies, which are in accordance with Euclidean axioms. In his *Elements*, Euclid proved that there are only five regular polyhedrons. Plato used all of them.

world, revealing an understandable, visually oriented longing for beauty. Even Plato investigated things primarily because he had the feeling that he would find an appealing form of beauty. That we no longer regard the smallest particles of matter, the atoms or their components, as things per se, but encounter them as symmetrical forms—as Plato had guessed—is discussed in a later chapter.

As we know, Plato was more active philosophically than geometrically, and thus he did give some thought to the question of what beauty is in his dialogues. Ultimately it was none other than Socrates who asked about beauty in his conversation with a so-called teacher of wisdom, the Sophist Hippias. In the dialogue Plato presents, Socrates makes it clear to his conversational partner that for many of us, it may be easy to say what is beautiful, for example, a beautiful urn, a beautiful horse, or a beautiful girl, but it is not as easy to say what beauty is, what all the objects and persons that we consider beautiful have in common and how they impress us.

Socrates argues in the course of conversation that beauty can be defined as neither the useful nor the proper and unfortunately also not that which is pleasing to the senses. For we feel pleasure, as Hippias points out, even when it has nothing at all to do with beauty. It becomes apparent in the course of the dialogue that there are no satisfactory answers to the question of what beauty is.

The Platonic line of argument is not completely useless, however, because Socrates's observation that "the most beautiful ape is ugly compared with the human face," but "the most beautiful girl is ugly compared with a goddess" is especially interesting for our purposes. Apes, humans, gods: The words Plato puts in Socrates' mouth that make philosophy teachers wink and nod seem reasonable and convincing, at least at first glance. When we give them a second read, however, we discover that Socrates brings two levels into play and mixes them so in the end it's impossible for us to draw any conclusions.

Are we not to take the quoted dialogue with Hippias, the teacher of wisdom, seriously? Was Socrates conducting an exercise in thought meant for the student of the Athenian academy rather than an investigation into the beautiful?

If we could raise an objection with Socrates, taking all these questions into consideration, we might say, "When you, O Socrates, compare apes and people aesthetically—in their sensory effect on my perception—and presume that the better position is that of the species to which we both belong, then your lecture makes sense only from the human standpoint. Otherwise, sir, you're going headlong in the wrong direction. You might heed the observation your contemporary Zhuangzi in China once made. He put it this way, 'Men have their beauty queens, but when a fish sees such a beauty, it dives to the bottom, and when a bird sees her, it flies away. Who knows what is really beautiful on this earth?' "[5]

To illustrate in modern terms this Chinese insight that the Greek overlooked, we can imagine how Robert Redford would leave a bonobo cold in comparison with a like companion, and how the somewhat homely impression of Socrates himself, as portrayed in numerous statues of the philosopher, wouldn't concern the chimpanzee in the least.

In my opinion, Socrates explained nothing with his reference to apes; his remarks can be better explained in terms of evolution, which I discuss in another chapter. As far as the goddess, who is supposed to be more beautiful than a girl, is concerned, neither Socrates nor any other mortal has ever seen a woman of the divine world in person. The comparison is empty therefore in terms of beauty because Socrates's perception of people and his perception allowing him access to the divine world have very little to do with each other, especially in a culture that was not to make a likeness of its god. Socrates might have been referring to some kind of ideal image of a gorgeous girl, the way European men might conjure up their Helen. To define an actually existing beautiful face as "ugly," as Socrates does, doesn't show wisdom—just bad taste.

Inner Beauty

What Socrates and Plato said about beauty would be taken up again and interpreted a few centuries later by the philosopher Plotinus when he was lecturing in Rome. Plotinus tried to tinker less with great ideas and to devote more time to understanding the human perceptions involved in beauty. Beauty, he decided, is found in the realm of the face; it is also found in the realm of hearing, in the arrangement of words, and in the whole arena of music (melody and rhythm are also beautiful things); in addition, it is found when we advance upward from the realm of perception to beautiful pursuits, actions, conditions, sciences, and finally the beauty of the virtues. It remains to be seen whether there is an even higher realm of beauty.

Plotinus didn't put his efforts into finding an all-encompassing beauty for all things—a kind of platonic heavenly ideal. He turned his gaze around 180 degrees and offers—in contrast to and as fulfillment of the Pythagorean view—an internal definition of beauty. "Beauty is the shining through of the eternal brilliance of the 'One' through material appearance." [6]

That Plotinus consequently gave accurate explanations of beauty for scientific contexts or cognitive experiences was demonstrated by Kepler, who also spoke of a light, of a flash in the soul that appeared the moment the inner and outer images converged. When the fit was perfect, he was granted a special way of understanding a truth about the world. He could recognize it by the brilliance of its beauty.

The "Line of Beauty"

After Plotinus, there is a long philosophical gap which wasn't filled until the eighteenth century when it was filled with a vengeance. Kant the "all-crusher" finally formulated his thoughts on aesthetic judgment and determined the direction of German-speaking philosophy.

A sympathetic but not especially popular contemporary of Kant was Erasmus Darwin, the English doctor and natural scientist usually introduced as the grandfather of the world-famous Charles. The world is unmistakably indebted to Charles Darwin for the first precise version of a thought that wasn't yet available (or only in its initial stages with Kant). This thought—evolution—concerns us only as far as it concerns beauty. After all, it is not enough to postulate inner images of the soul or the idea of goodness, we also have to explain why the images appear and what form they have taken in the entire context of biological history ending with human beings in all their beauty.

I can only suggest this approach to evolutionary aesthetics and just begin to scratch the surface. At the moment I am less interested in focusing on Charles than on his grandfather Erasmus Darwin. Erasmus belonged to a separate rank from his grandchild. In his book that appeared in 1794 under the title Zoonomia, Erasmus attempted to discover the laws governing organic life. He didn't just think about what people found beautiful, he also wanted to know why they found a thing beautiful. In other words, Erasmus was already searching for refined evolutionary perspectives long before his grandchild started searching systematically and comprehensively.

The starting point of his earlier meditations is the statement attributed to the English painter and copperplate engraver William Hogarth that people find bodies and forms especially appealing when they flow in an s-line (Fig. 2.5). Hogarth talks about the "line of beauty" that he has found in much of nature and art and that people simply enjoy viewing (Fig. 2.6).

Erasmus Darwin knew just what Hogarth was discussing. For him, the line of beauty was also the s-curve, and he has given a marvelous reason, once known as the Erasmus Darwin hypothesis. In his book Zoonomia, Darwin suggested that aesthetic perception is determined by

Figure 2.5.   Table 1 from *The Analysis of Beauty* by William Hogarth. (1969) London: Scolar Press.

early sensory experiences, and he bases his suggestion on the experiences that most people have in early childhood:

> When the baby soon after it is born into this cold world, is applied to its mother's bosom; its sense of smell is delighted with the odour of her milk; then its taste is gratified by the flavour of it; afterwards the appetites of hunger and thirst afford pleasure by the possession of their objects, . . . and lastly, the sense of touch is delighted by the softness and smoothness of the milky fountain, the source of such variety of happiness.
>
> All of these various kinds of pleasures at length become associated with the form of the mother's breast; which the infant embraces with its hands, presses with its lips, and watches with his eyes and thus acquires more accurate ideas of the form of its mother's bosom, . . . And hence at our maturer years, when any object of vision is presented to us, which by its waving spiral lines bears any similitude to the form of the female bosom whether it be found in a landscape with soft gradations of rising and descending surface or in the form of some antique vases, . . . we feel a general glow of delight, which seems to influence all our senses; and, if the object be not too large, we experience attraction to embrace it with our arms and to salute it with our lips, as we did in our early infancy the bosom of our mother, and thus we find, according to the ingenious idea of Hogarth, that the waving lines of beauty were originally taken from the temple of Venus.[7]

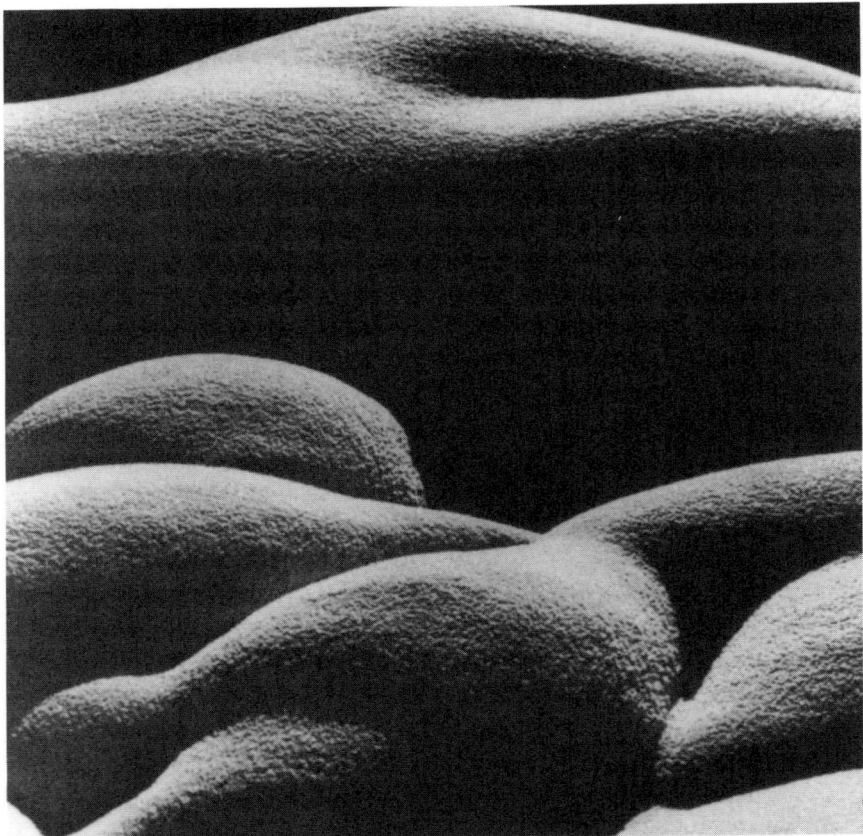

Figure 2.6.   A photograph by Ernst Haas from the book *The Creation*, Viking Press, New York, 1971.

Erasmus Darwin wrote about women at length, and it must have been particularly satisfying to him to have reached some biological or scientific insights. (Even Casanova contributed important evidence toward the biology of beauty—though nothing to be applied pedagogically.) Unfortunately, Erasmus didn't develop his thoughts on aesthetic consciousness any further, and so the question remains regarding how the wavy snake line of beauty disengaged from its sensory origins and became abstract, how it shed its individual skin and entered the collective head of the observer.

A wavy line that brings delight or a snake that frightens—that both figures are implied in the same form gives an idea of the difficulty of answering these questions. In any case, human perception clearly centers around the curvature Hogarth analyzed and practiced as the "line of beauty."

A Theory of Sensory Cognition

Directly after the English endeavor to explain sensory consciousness follows the German concern for defining sensory cognition. Immanuel Kant was the first philosopher to attempt a systematic theory of beauty in the *Critique of Judgment*, a work I have already quoted briefly. In this text Kant also wrote the last work of philosophy in which nature is mentioned as an equal to art in terms of beauty. The beauty of nature for Kant is just as self-evident as the beauty of art, but this thought unfortunately ends with him. After 1800, philosophers decided that nature could be put aside while they grappled with questions of aesthetics; as a result, their usage of this word became more and more distant from the Greek root *aisthesis* (perception).

During a brief period before Kant, everything was different. Another German philosopher had even taken a special approach in order to start up an independent theory of the sensory cognition which he called *aesthetics*. I am referring to the today essentially unknown Alexander Gottlieb Baumgarten (1714–1762) who taught in Frankfurt (on the Oder), wrote in Latin, and laid down an *Aesthetica* after 1750 in two comprehensive volumes, which was to be a theory of sensory cognition.[8]

In these books, Baumgarten worked on a reevaluation of the sensory access to things because he considered the assumption that there were mathematical and logical structures that would succeed in grasping the wealth of perceivable phenomena to be a fateful error. The hope of explaining all things via numbers stemmed from the days of the scientific revolution at the beginning of the seventeenth century when Galileo Galilei and René Descartes stated that the book of nature is written in a mathematical language. According to this school of thought, we learn to read the book of nature when the world is made measurable and can be understood through numbers.

Baumgarten admired the science of this doctrine, but he wanted to give precedence to the insights available to humans through the senses over the theoretical knowledge of science. According to his view, "the senses are surer than all other cognitive abilities." Tommaso Campanella expressed it the same way in his 1620 work *Del senso delle cose e della magia*.

Baumgarten drafted a science of sensory cognition (*scientia cognitionis sensitivae*) in which he outlines a form of expression consisting of concrete images instead of abstract lines of argument, and he called it *Aesthetics*. The goal of his aesthetics was to exercise sensory cognition perfectly, and this perfection of the sensory—the *perfectio cognitionis sensitivae*—was beauty for Baumgarten. Under this aesthetic umbrella, when we recognize beauty, beauty itself becomes a sign that we have gained a piece of knowledge through our senses. According to Baumgarten, beauty cannot be understood objectively, as a characteristic of the perceived object, and it cannot be understood subjectively simply as a pleasant feeling. Baumgarten tried to overcome this Cartesian split by creating unity again and finding the harmony that captures people's interest.

Baumgarten referred to the idea of unity in the broad sense, motivated by his insight—which remains contemporary—that in the course of scientific development, traditionally opposing domains actually belong together. Copernicus introduced this conflict when he placed the sun in the center and the earth circling around it. When our planet rotates, no matter how it looks to us, the sun doesn't go up or down. To say that the evening sun in the west submerges into the ocean is objectively false but is in accordance with the laws of the directly observable. In the words of the modern philosopher Hans Blumenberg, "Four hundred years after Copernicus, we still let the sun go up and down and the light of the moon shine down on us even though we certainly 'know' the objective relationship, and no one thinks to protest."[9]

In the seventeenth century, Kepler called attention to the conflict that existed between what we call aesthetic and what we call objective truth. In his *Astronomia Pars Optica*, he pointed out the difficulty people have distinguishing their words from "their eyesight."[10] Kepler tried, despite this, to reconcile literary and scientific theorems with each other—a task posed in our day only very recently.

Each single solution falters when it comes up against the distinction suggested by Descartes, which found enthusiastic agreement at the time. Descartes the rationalist abandoned sensory perception and only paid attention to mathematical information about the world. He drove the senses out of science and robbed it of its heart. Baumgarten, on the other hand, made the first attempt, with his aesthetics, to unite rational and sensory cognition. One of his points that we should take more seriously today concerns the aesthetic character of knowledge and states that a piece of knowledge is always something we express; in other words, we give it a form and use this form to present the knowledge to others, and the form is every bit as important as the content poured into it. Perhaps the question of form presents the best place to start to give aesthetics in science the status it deserves.

An Inquiry Concerning Beauty

Even before Baumgarten, the Scottish philosopher Francis Hutcheson presented an inquiry concerning "beauty, order, harmony, design" in which he also had in mind science and its mathematical clutches on the world.[11] In his 1725 treatise, Hutcheson spoke in one chapter expressly about the "beauty of theorems." It didn't take him long to try to define beauty in general, and he defined it in a new way so subjective and objective elements come together.

Next, Hutcheson clarified the quality defined by the word *beauty* and emphasized that people have a sense for it. He wrote, "The word *beauty* is taken for *the idea raised in us,* and a *sense* of beauty for *our power of receiving this idea."*

After this opening phrase, Hutcheson devotes himself to the question of how we can distinguish the external objects that evoke in us the idea of beauty, and at the same time he suggested that one of the signs is "uniformity amongst variety." We can find this quality in nature as well as in art. In addition to concrete objects, it also allows intellectual constructs such as mathematical theories to be beautiful.

Hutcheson believed that the practice of science offers a special opportunity to discover uniformity among variety, and he introduced three levels where the researcher has access to beauty.

The first level is for appearances that we can perceive directly. These appearances, the clouds we see in the sky every day or the starry sky we gaze into at night, build the foundation for all scientific investigation. No one needs to be told what is meant by uniformity or variety here, and it doesn't take a theory for us to know something by perceiving its appearance or to feel that we are seeing or recognizing beauty.

Scientific method made its entrance onto this level a long time ago but the wealth of this level has yet to be exhausted. After all, there are still astronomers who are impressed by celestial spheres even when they can't observe them directly and have to look at them in the pictures made by modern devices in the cosmos (Fig. 2.7). Similarly, there are chemists who rave about the beauty of the molecules with which they work (Fig. 2.8) although these structures, too, are only accessible with special equipment and can't be perceived with the naked eye.

To be exact, we should say that chemists find the model they have "built" beautiful rather than the molecule itself. The world of the model takes us to the second level, where Hutcheson saw potential beauty from the smallest molecule to the largest universe. Ptolemy built a model, for example, and Copernicus and Kepler made it more beautiful by changing its meaning.

From the first to the second level there is an important difference in terms of beauty. Those outside the scientific realm rarely consider models formed from theories to be things of beauty. In general, an astronomer will find it easier to discover more regularity (uniformity) within planetary motions among the bewildering variety than a layman will, and the same holds true for the model of the molecule. A chemist can see the variety of functions encompassed in the uniformity of its structure more comprehensively, and only a chemist knows in the end how much work is hiding in the beautiful form that everyone likes so much.

The second level of the aesthetic appraisal of science is where the concept of evolution comes in, the idea that found its way in the nineteenth century into the heads of natural scientists, and mainly into Charles Darwin's. Darwin even had an expressly aesthetic sensation at the thought of imagining the whole of nature as a close network of integrated creatures determined by one force—natural selection. Natural selection embodies uniformity in natural variety and this lends evolutionary thought beauty.

The third and highest level, as Hutcheson envisioned it, is the level of abstraction where theorems and theories are found. These are beautiful when they have a form with the highest degree of simplicity—uniformity— and cover the most territory—variety. A theory is thus beautiful when it is presented as a principle that we can understand effortlessly and from which we can draw far-reaching conclusions. An example of this would be Einstein's theory of relativity, which in its special version needs only the constancy of the speed of light and the equivalence of the inertial system to give rise to a new and surprising understanding of space and time.

Figure 2.7.   A beautiful galaxy in the form of a spiral; we have been aware of this modern image of the sky since the nineteenth century.

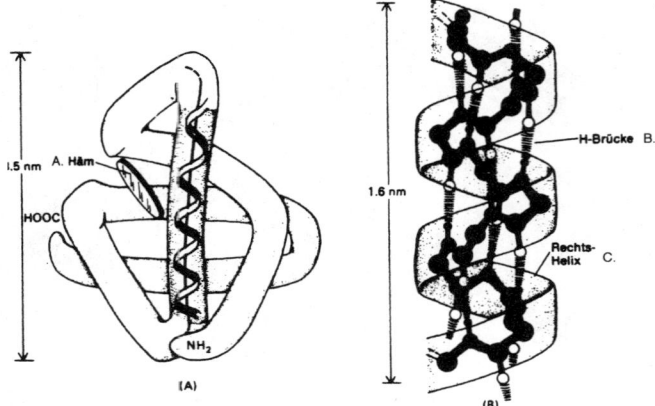

Figure 2.8.   Some molecules from the reservoir of biochemistry. Depicted is (A) the line of the alpha helix, devised in our century at the beginning of the 1950s by Linus Pauling even before experiments reached this degree of accuracy. The helix shows the way in which the elementary components of protein can build a chain. Proteins function as molecular catalysts in cells and enable their life-sustaining reactions. One of the (archetypal) basic components found in its structure is the alpha helix, as in the well-known example of hemoglobin, whose job is to absorb and transport oxygen in blood. A. Heme; B. H bond; C. Right helix.

Another example would be the principle of least action that makes the path that a fragmented ray of light travels as easily understandable as what makes the beautiful sport of figure skating possible—the melting of ice under pressure. The principle of least action, unfortunately somewhat forgotten today, can reduce many phenomena of the physical world into a simple and thus beautiful thought. At the beginning of the twentieth century, its property was still at the center of many scientific analyses, and even Max Planck, whose formulations were generally otherwise somewhat modest and reserved, raved when he started to talk about the "highest physical laws." He called the principle of least action "the crown" of physics, because it contained the "four world coordinates in perfectly symmetrical arrangement." [12] Four principles of equal value radiate symmetrically from this central principle corresponding to the four world dimensions; the spatial dimension corresponds to the (threefold) principle of momentum; the time dimension corresponds to the principle of energy. Never before was it possible to follow the deeper meaning and the common origin of these principles so far back to their roots.

It's obvious that Planck talked about the principle of least action the same way Plato talked about his regular solids. Whereas the geometrical images are easy to present, the dynamic principle stubbornly resists all simplification. Mathematically, it calls for a time integral that assumes a minimum value and physically, that change (motion) in nature allows as little as possible expenditure. The principle, first proposed and then neatly formulated in the eighteenth century, earned a special meaning when it was found to function with systems in which the details of the mechanism were unknown. Whether fluidity rotates or balls roll, whether states of aggregation change (the melting of ice) or temperatures equalize could always be indicated by the principle of least action occurring in any given system. For Planck and other physicists, the wonderful thing about this principle was that the variety of the physical disciplines—thermodynamics, mechanics, electrodynamics, optics, and many more—could be uniformly understood. Planck presented the "uniformity amongst variety" now recognized to be beautiful, as the goal of natural science. These thoughts, which he was continuously emphasizing, he articulated for the first time in a famous lecture on the uniformity of physical theories of life ("Einheit der physikalischen Weltbilder") in 1908: "Ever since the observation of nature has existed, it has held a vague notion of its ultimate goal as the composition of the colorful multiplicity of phenomena in a uniform system, where possible, in a single formula."[13]

In other words, the goal of science was beauty from the outset, and at the beginning of the twentieth century we still knew this. Perhaps we will learn this once more at its end.

Beauty According to Leibniz

The direction aesthetic theories took after the eighteenth century— that is after Hutcheson and Baumgarten—had increasingly less to do with science, and thus we leave off here. As preparation for the upcoming chapter on mathematics, I should mention Gottfried Wilhelm Leibniz, who in the seventeenth century found beauty in his mathematical–physical search for knowledge.[14]

Leibniz spoke with great pathos about beauty in science: "The beauty of nature is so great and its contemplation so sweet . . . whoever tastes it, can't help but view all other amusements as inferior."[15] Leibniz even spoke about the desire for beautiful theories, and among other things, he

In his analysis of beauty, Leibniz observed that the sensation of beauty in science can change over the course of history. What an earlier generation considered beautiful, he declared, can seem trivial and banal to the next. Beauty can become dated, but Leibniz didn't stop at this negative insight. He also realized that the quality most beautiful theoretical discoveries share is appearing easy and self-evident. For Leibniz, who recognized beauty in nature and theories alike, "simplicity in variety" (Hutcheson) meant the harmony of things and their parts, in short, beauty. According to Leibniz, men and women should seek out beautiful truths, because when they do, they serve as a mirror of God, who created "the best of all worlds" in perfect beauty.

If we're only concerned with the beauty of a result, as Leibniz discovered with his own attempts at mathematical theorems, we stand the chance of going wrong. The pursuit of beauty can fail, and the history of natural science is full of cases like this. Leibniz saw the difficulties here for natural scientists who believe and put their trust in the thought of an all-encompassing perfection and beauty for all forms of life. However, difficulties are meant to be overcome, and that is entirely up to the individual.

3   *The Irrational Measure of Things:*
The Long and Short of the Golden Section

Beauty is the first test: there is no permanent place in the world for ugly mathematics.[1]

Godfrey H. Hardy

In the nineteenth century, there was a phase in which psychological researchers tried to make their study of beauty an exact science as had been done in physics a few hundred years before, by taking everything that seemed aesthetically interesting and considered beautiful according to conventional taste and making it measurable and countable.[2] They counted words in well-known poems and heads in famous paintings, determined whether sculptured figures were looking more toward the left or more toward the right, and tried to establish what could be understood more quantitatively in the fine arts.

The narrow technical focus of these researchers, who were otherwise unconcerned with their subjects, soon led them to such basic insights as the realization that rhyme at the end of a line of poetry is essentially for the enjoyment of those who read poems, and that poetry is especially suited for declamatory elocution; its lines demanding only about 3 seconds. The latter, in no way trivial, observation applies beyond cultural borders, characterizing European poetry as much as Japanese or Arabic. This meter is found in Goethe's poem "Holy Longing" from the *Westöstlichen Diwan* (West–East Divan) whereby for two lines the poem sounds like a kind of warning:

Don't tell any but the wise
For the masses will just mock
I praise what is truly alive
that longs to die in flames.*

The Golden Rectangle

Gustav Theodor Fechner, one of the nineteenth-century psychologists who took an interest in physics, set out to measure beauty with stopwatch and tape measure. Carrying out his project involved, among other things, visiting museums and measuring the frames of hundreds of masterpieces. The proportions of the average frame, he discovered, were pleasing. According to his measurements, in most cases the dimensions of the frames qualified them as golden rectangles (Fig. 3.1).[3] A golden rectangle is defined as a rectangle in which the ratio of the shorter side (a) to the longer side (b) is the same as the ratio of the longer side to the sum of both sides ($a + b$).†

Back in the psychology lab, Fechner followed up his museum visits by presenting different types of rectangles to research subjects and requesting that they tell him spontaneously which shape they liked best. Not surprisingly, the majority of them chose the golden rectangle. No other four-cornered shape, even the square, has ever found as much favor as the golden rectangle. If we consider printed material, for example, we find that the layout of books and graphic design more often than not follow the form of the golden rectangle (Fig. 3.2). But why do we find the proportions of the golden rectangle pleasing, and how do we explain this preference?

The attribute "golden" comes from a work of ancient geometry that discusses dividing a given side so the ratio of the two parts—the length of the whole to the length of the larger section and the length of the larger section to that of the smaller section—is the same (Fig. 3.3). This division became known as the "golden section," or *sectio aurea* in Latin, and the basis of its aesthetic appeal was readily apparent; because each individual section was comparable to the next, every section was reflected in the other, or to borrow a term from chaos theory, *self-similar*.[4]

---

*Sagt es niemand, nur den Weisen/Weil die Menge gleich verhöhnet,/Das Lebend'ge will ich preisen,/Das nach Flammentod sich sehnet ("Selige Sehnsucht").

†I'd like to add to my relatively unsystematic observations that Fechner was probably right about the "golden mean" as far as the usual pictures in museums go. Well-known paintings, however, those by Vermeer or van Gogh for example, tend to stray from these aesthetic norms. Perhaps this is what makes them masterpieces.

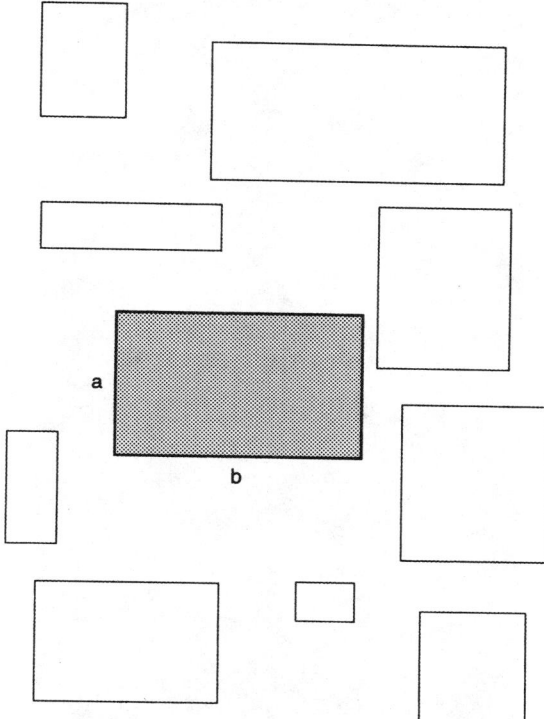

Figure 3.1.   Various rectangles, one of which (with sides a and b marked) qualifies as "golden."

Greek architects were already using the golden section in the design of the Parthenon (Fig. 3.4), and during the Renaissance, architects used the golden section to create aesthetically appealing surface patterns and to determine distances. A notable example is Filippo Brunelleschi's Capella Pazzi in Florence.

For some Renaissance artists, the golden section was the "divine proportion" of the book *De divina proportione* written by Fra Luca Pacioli in 1509. There is also evidence that this "holy ratio" was known and valued even earlier in Egypt. The Great Pyramid of Giza's ratios of the heights and lengths of its sides basically correspond to the numerical value of the golden section.

If we were to measure the front of the Parthenon as accurately as possible and determine the ratio of the shorter side to the longer one, we

Figure 3.2.   A book in golden rectangle format.

would find something like the quotient 1 : 1.618 (keeping in mind that each person measuring would come up with a different number, if the starting and ending points to be measured were not defined). The golden number—or the golden section's value—can be found most easily if we let the length of the side to be divided be 1 and the longer side be G (standing for "golden"). The shorter section then has the length $(1 - G)$, and the *sectio aurea* is defined by the equation $1/G = G/(1 - G)$. Another way of stating this would be through the quadratic equation, $G^2 + G - 1 = 0$, which can also be written:

$$G = \tfrac{1}{2}(5^{1/2} - 1)$$

At first glance it may seem strange for this to be the golden number or the golden ratio, considering the claims to its divinity and golden beauty. If we

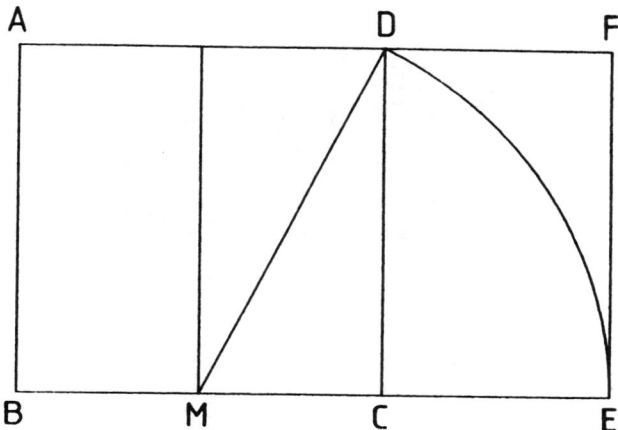

Figure 3.3.   The construction of the golden section according to Euclid. The starting point is a square (ABCD). The middle (M) is determined on one side and the
corresponding point is connected diagonally to the vertex. If a circle with the
radius MD were drawn around M, it would cut off the extension of the side with M
to form a new point, E. Thus a new, longer section BE is created, which is divided
by C in the ratio of the golden section (only asserted but not proven here).

do the math again, worked out to six decimal places, G is approximately
0.618033.

If you're surprised that the same three numerals fall after the decimal
point in the Parthenon ratio, you'll notice that the golden number is closely
related to its reciprocal and can be arrived at by adding the number 1.

Because we can start from either end when coming up with the golden
ratio—either we begin with the length of the whole or with the length of
the smallest side—we can also take the golden number as defined previously, find its reciprocal, and call this the golden number. What falls to the
right of the decimal is always the same, and whether a 1 or a 0 stands in
front of the decimal is up to whoever is taking the measurements (which
sometimes makes the writing on this subject confusing). If we have a 1 in
front of the decimal—if the golden number begins with 1.618—we are led
to the pleasing discovery that we can get the reciprocal of the golden
number by subtracting 1, and the square by adding 1, and so on.

These are only some of the games played with the golden number.
What I find significant is that in this number many see the key not only to
beauty but also to life.

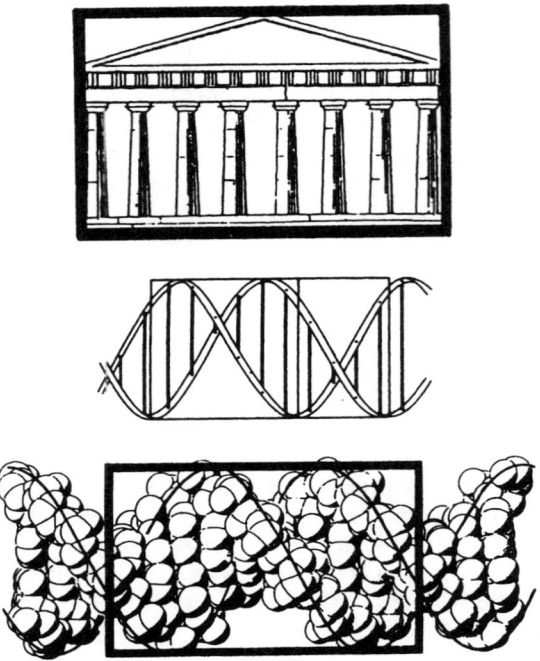

Figure 3.4. The Parthenon and the double helix comply with the specifications of the golden section. This must be one reason both are considered classics.

The Idea of Procreation

The question is: What is so special about this number, which actually seems ugly at first? What is beautiful about $\frac{1}{2}(5^{1/2} - 1)$? And what does this strange expression and its numerical value have to do with life?

In fact, the number itself seems a bit of unpleasant, and it's not easy to see the harmony. Kepler even called it the "inexpressible" number, by which he meant only that G is not a rational number, as the mathematicians say. Ultimately the golden ratio lurks in the root, and it cannot appear as the quotient of two natural numbers.

In today's terminology, G is therefore an irrational entity and is correctly defined in terms of science as a number that can't be written as a ratio of two whole numbers—as many people find *irrational* an irritating word.

In the seventeenth century, this expression of the irrational within apparently pure, rational mathematics rekindled the unpleasant association of nonsense, and Kepler avoided it by using the concept of an inexpressible number instead. He found the properties of G wonderful, even calling G "divine," enthusiastically referring to it as the "the jewel of geometry." Kepler raved about its "wonderful nature and its many special properties, the most important one—that you can continuously get a section divided into identical parts. If you add the larger part to the whole, and the larger part to the smaller, then what had been the whole becomes the larger part" (Fig. 3.5).[5]

Kepler elaborated accordingly in his main work *Harmonices mundi* and further developed his treatise on the golden ratio and its inherent self-similarity: "This beautiful ratio conceals the idea of begetting. For as the father begets the son, and the son another, each one similar to him, so it is with each division of the ratio when the larger section is added to the whole."[6]

If Kepler was right, if this means that the golden section is a crystallization of the dynamics of life processes and that it actually shapes growth and reproduction, then it's easy to see the reason for G and the golden division's beauty. The golden section is then an indication that life is successful and on its way to reproduce more life. Such a view must agree with one's perception and please and benefit the one perceiving. (In any case, it would explain why, in the context of human evolution, people are known to get a good feeling when they perceive the golden section.)

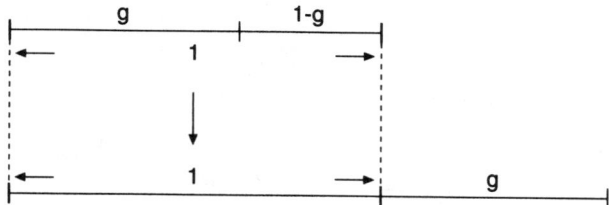

Figure 3.5.   The proper extension of the golden section past the larger part produces a new side that is again divided according to the golden ratio. This can be easily calculated because the old proportions, $1/G = G/(1 - G)$, lead to the new proportions, $(1 + G)/1 = 1/G$, which lead directly to the same quadratic equation, meaning back to $G^2 + G = 1$.

The Multiplying of Rabbits

A human figure (Fig. 3.6) makes a pleasant impression on the senses when it is drawn so it reveals the qualities of the golden section. Leonardo da Vinci produced his famous proportional sketches around 1490 according to the specifications of the classical architect Vitruvius to illustrate a harmony of parts among each other and in respect to the whole, like the arrangement of beauty required in the sense of Pythagoras.

But do people really look like that? Has nature really built us so, for example, the navel divides our bodies into two halves, either 1.618 or 0.618? Anyone who starts measuring friends and relatives will find the results sobering. However, we can by no means draw the conclusion that most people are not as beautiful as we would wish, for the sculptures praised as the classic ideal, such as the Apollo of Belvedere in the Vatican museum, are clearly missing the figures 618 after the decimal point.

As it turns out, nature appears to devote a greater amount of time to observing more laws of growth than attending to and fulfilling the golden section, and beauty perhaps reveals itself more in deviations from the ideal

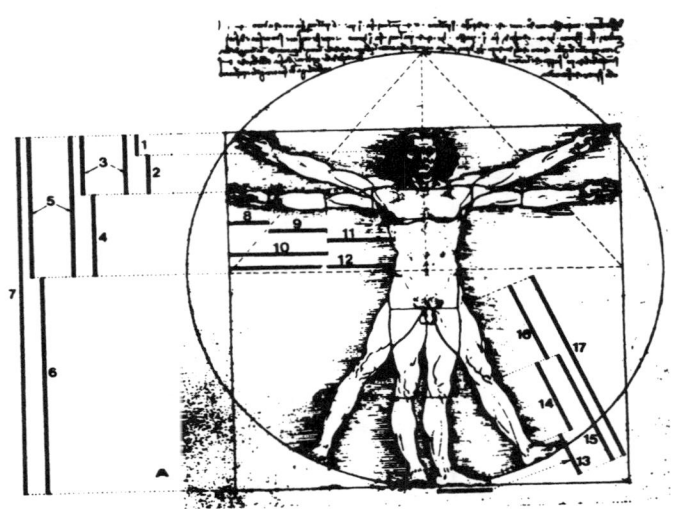

Figure 3.6.   The proportional sketches by Leonardo da Vinci made around 1490 and fashioned after a model of the ancient Roman architect Vitruvius. Vitruvius wanted to use human measurements for the foundation of construction. Leonardo's figure is standing as if it were imitating a pentagram. The golden section can be seen everywhere (cf. Fig. 3.7).

in the right measure than in predetermined perfection. Still, Kepler's persuasive and clarifying concept should be effective when the growth processes are simple. Another Leonardo concentrated on just this kind of situation around 1200, about 400 years before Kepler. His name was Leonardo of Pisa, better known by his nickname Fibonacci. Mathematicians are well acquainted with the Fibonacci numbers that are generated by a simple formula. The series begins with the numeral 1 and each subsequent number is the sum of both preceding numbers (with the addition of the imaginary 0) as follows:

At the outset is:

$$1, 0 + 1 = 1; 1 + 1 = 2; 1 + 2 = 3; 2 + 3 = 5; 3 + 5 = 8; 5 + 8 = 13; 8 + 13 = 21,\text{ and}$$
so forth,

which implies that the whole series continues:

1, 1, 2, 3, 5, 8, 13, 21, 34, 55, 89, 144, 233, 377, 610, 987, and so forth,

which does not look particularly beautiful, at least not when we are looking behind the "mathematical scenes." The series expands very quickly, almost to the point of exploding, and this is precisely what interested Fibonacci, alias Leonardo of Pisa. When he thought up this set of provisions, he was hoping to get a firm mathematical grasp on the reproduction of rabbits (his calculations do not account for the deaths of the animals).

Fibonacci wanted his rabbit reproduction calculations to approximate the biological conditions, and in his closing thoughts he distinguished between two kinds of animals, between the rabbits capable of producing offspring, and the reproductively immature rabbits, just entering the world. Fibonacci wondered how many rabbits a single pair, itself excluded from the calculations, could produce in the long term if in a typical given time span a new pair came into the world and in the same span later this same pair became capable of reproduction and acted accordingly.

If we try to convey this in sober mathematical formulas, we see that with these givens, the population increases just as the series a la Fibonacci indicates. Kepler was familiar with this series and had many thoughts on the subject.

First, the fractions of consecutive pairs—that is ½, ⅔, ⅗, ⅝—form part of the seven harmonies that Kepler as a scholar of music found particularly splendid. They correspond musically to the tonal intervals of octave, fifth, and major and minor sixths.

Second, the series of fractions eventually reaches a maximum limit, and this limiting value is none other than the golden ratio. If we call the nth Fibonacci number F_n and the next value F_{n+1}, for example, $F_4 = 2$ and $F_5 = 3$, then as Kepler calculated, and as anyone interested can easily check, we have:

$$F_n / F_{n+1} \rightarrow G$$

Expressed in words: For increasingly larger Fibonacci numbers, the quotient of two neighboring numbers approaches the value defined by the golden section.

This result allows us to make a hypothesis: Whenever a form of growth exists in nature distinguished by Fibonacci numbers, we can expect to find the golden ratio. Additionally, when this ratio determines the measure of growth, then nature generates the forms that please us. These forms appear in the beauty of the mussel (Fig. 3.7) whose growth spirals can be clearly designated as Fibonacci numbers.

Now the principle of growth shows up in the spirals according to the golden section (Fig. 3.8), which we can quickly verify by looking at the blossoms of a sunflower or carline thistle (Fig. 3.9), or by counting scales of fir cones or by determining the inflorescence of Compositae, the sunflower family. The natural scientists Peter H. Richter and Hans-Joachim Scholz have presented this finding in detail as follows:

> The spiraling arrangement of leaves in phyllotaxis is probably the most spectacular example of the golden section as realized in nature. Whether it is the needles on a young pine bough, the scales of a fir cone, the leaves

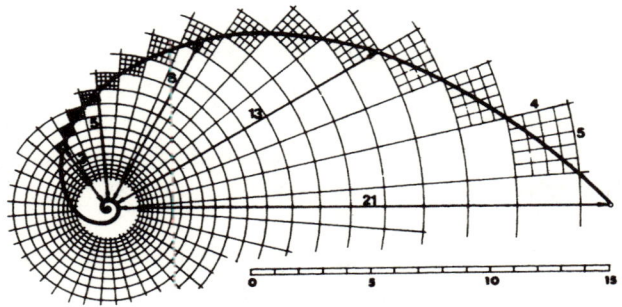

Figure 3.7.   The spiral of a mussel shows how the golden section is related to growth. [From Cramer, F., & Kaempfer, W. (1992). *Die Natur der Schönheit* (The nature of beauty). Frankfurt: Insel.]

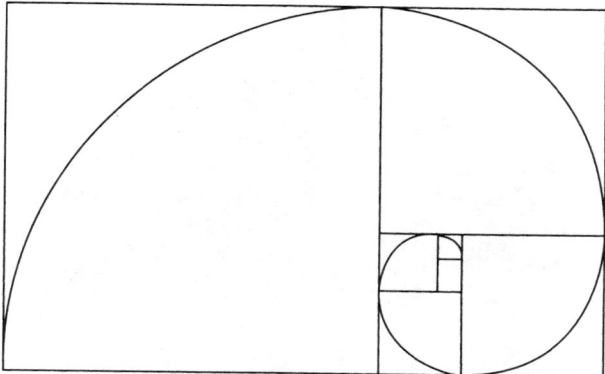

Figure 3.8.   We can draw a golden angle (see Fig. 3.3) and divide it into a square and a second (smaller and faced in a different direction) golden rectangle. The process can be repeated and continued *ad infinitum*. In this arrangement of increasingly smaller squares and golden rectangles, it is now possible to draw a spiral as a series of quadrants in the resulting squares that move around the point in which the diagonals of all the golden rectangles would meet (intersect). The spiral possesses the property known as self-similarity, which means that we can take a portion of it and find it similar to the whole. Once again we find that the idea of reproduction is inherent in this process and so it is no wonder that we find spirals so often in nature (see Figs. 3.7 and 3.10). Self-similarity thus leads to beauty.

of a cabbage head or the inflorescence of Compositae, the angle measured from the axis between two successive developing leaves (needle, scale, leaf, etc.) is almost always, providing the availability of good nutrition, the figure 137.5°. This *golden angle* divides the circumference in the golden ratio:

$$(1 - G) \times 360 = 137.5$$

because in the case of cabbage or of the artichoke, and in the branching of young oaks or the leaf blossoms of a rose, we can easily recognize this angle in that as it progresses from one leaf to the third next, the axis spirals a little more than once around: $3 \times 137.5 = 412.5 = 360 + 52.5$; the fourth leaf forms an angle of a little more than 50° with the first, which, with some practice can be easily done.

Still more impressive however is the appearance of the *Fibonacci spirals* as a consequence of the golden angle [Fig. 3.10]. We can see the connecting lines between next-door neighbors in the lattice blossoms that spiral clockwise or spiral counterclockwise. The number of the spirals spiraling clockwise and the number spiraling counterclockwise are different. Not counting exceptions, the two numbers build a pair of neighboring Fibonacci numbers: according to how densely packed the

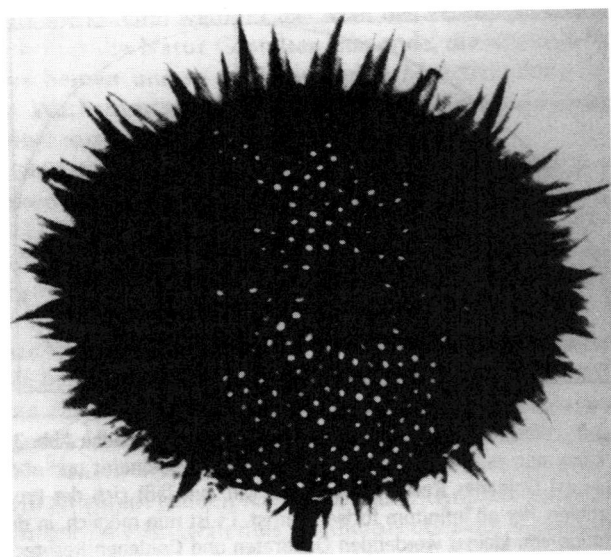

Figure 3.9.   The Fibonacci series can be seen in the spirals in the inflorescence of a sunflower.

leaves are radially or in height, we count 3 to 5 spirals (with pine cones), 5 to 8 (with fir cones), 8 to 13 (with pineapple fruit), 13 to 21 (with daisies), 21 to 34, or even 34 to 55 (with sunflowers, thistles, and the like).[7]

The Most Irrational Number

Anyone who looks around at nature knows that many plants look much different from what the golden section and its spirals suggest. With

Figure 3.10.   More examples of spirals that look beautiful by following the golden laws.

many papilionaceous flowers the angle between consecutive leaves is 180°
or 90°, as with Labiatae. However, if we look a little more closely, we
suddenly find that there is no third possibility between these "round" or
whole numbers and the strange (irrational) golden angle. Nature appears
choosy: It is either golden or whole; in this case, either right angled or
leaning forward.

The question emerging at this point is, What distinguishes the cir-
cuitous golden number not only from the round and even values, but also
from all potential figures? The answer is in the word that Kepler wanted to
avoid, which we now, however, have to use at full strength. The golden
number is not just an irrational number among many—it is the ultimate
irrational number (even when not immediately apparent). No other number could be more irrational than the golden number!

In their language, mathematicians can accurately prove this equation
by defining rules for determining whether a number is more or less irrational. The proof involves arithmetic as well as how well we can approach an irrational number by way of rational numbers. We can take on
each irrational number, for example, pi, the famous ratio of circumference
and radius, with fractions from whole numbers, for example, pi with 22/7
or 333/106, but the question is, in the best-case scenario, how close can
we get?

The golden section has the special feature that it can be written as a
rational number, thus as the fraction $G = 1/(1 + G)$, but it slips through the
back door a second time. If we write the fraction again so that G is in the
denominator, we get $G = 1/[1 + (1/1 + G)]$, again an equation that can be
started over from the beginning. If we continue on our way, eventually we
get what mathematicians call a continued fraction:

$$G = \cfrac{1}{1 + \cfrac{1}{1 + \cfrac{1}{1 + \cfrac{1}{1 + \cdots}}}}$$

The model is easy to construct and frankly speaking, a waste of paper. Its
most important aspect is that over and over again it is the numeral 1 that
begins the next set in the denominator. This is a decisive pattern because in
terms of strict mathematics it can be generally proven that an irrational
number is poorly approximated by rational numbers if very small numbers appear in the denominator when it is presented as a continuous

fraction. No smaller number than 1 can appear there, and 1s subsequently emerge with the golden number as shown in the illustration. It is thus maximally irrational, and this property, now more than ever, raises the question why it appears beautiful to us in any particular form. What put irrationals on their way to beauty?

Stability in Perturbation

Perhaps the answer is in the heavens, and our gaze should have turned in that direction a long time ago. From time immemorial, as Kepler formulated in the seventeenth century, observers of the cosmos and the planets have surmised that "the motions of the heavens are nothing more than a continuing chorus of many parts made comprehensible through reason."[8] If musical intervals are hidden in the Fibonacci numbers, shouldn't the golden section be found somewhere in the (revolutionary) orbit of the planets or anywhere else in the universe?

The answer is fundamentally "yes"—as to be expected—but the way the golden section's irrational nature leaves behind its impression is surprising. To imagine it and be able to understand it, we have to take a small trip into history, beginning with Newton. Newton, as we know, discovered that the physical laws on earth are the same as those in the heavens. Beyond this he was in the position to issue the laws of planetary motion found by Kepler through year-long, painstaking observation of the formulaically understood force with which two masses mutually attract.

According to Newton's laws, the universe functions like a big clockwork that the Lord God once made before prehistory and set into motion. People at the time tended to see in this imagery the bondage that He created for them, but in the end it gave them the beautiful feeling of eternal stability and for what more could they ask? In the cosmic clockwork the small wheels of the planets interlock precisely, forever, and for all time. They could rely on the Lord.

Or so they thought until, in the course of time, it gradually became clear that Newton's motion equations only secured stability as long as only two heavenly bodies were observed at any given time, such as the motion of the planet Mars around the sun or the turning of the moon around the earth. If we were to add a third object to the heavenly bodies, we could still set up the corresponding Newtonian equation, but we could no longer find an exact solution. It had already failed to work for the three-body problem and went seriously wrong with even more difficult many-body problems. There was no solution for these and thus no fixed or predictable orbit. They could only make approximations, and as a result, it became increasingly apparent that Newtonian mechanics were no guarantee for a stable world order. On the contrary! The physicists discovered the ever-greater likelihood that disturbing external influences, such as comets, could shake up even the most apparently stable orbit and, in fact, without direct impact. A comet, for instance, could create a disturbance simply by putting its mass, and thus new forces, into motion.

In the nineteenth century, the question regarding the conditions for the stability of the solar system became increasingly urgent. In 1885, Swedish King Oscar II offered a prize for the scientist who could come up with an answer and Frenchman Henri Poincaré won the prize. Poincaré didn't solve the problem to general satisfaction, but he did give the investigation the deciding direction by drawing attention to the question of the stability of planetary motion subjected to perturbation. To be more precise, Poincaré concentrated on the three-body problem and so observed two planets orbiting a star cluster each with its own frequency. As an example, we might imagine Saturn and Jupiter on their way around the sun.

Poincaré could, for one, show in the context of his *New Methods of Celestial Mechanics* that a system's behavior, its stability, is determined by the orbit that stands in some form of resonance with an incoming perturbation, a comet for example. He could even determine how likely the chances were that one of the two circling planets might spin off in an unstable motion and abandon the system that has broken up. The probability was dependent on the ratio of the frequency assigned to the orbits or periods of both planets. Poincaré discovered to his surprise that even the weakest perturbation could overthrow a solar system and could break apart the single orbit if the frequency relationship were a rational number.

Poincaré made great progress but he never took the most important step. That this step was taken only in the 1960s and that it took the meeting of several mathematicians shows just how difficult the step was. In 1963, Vladimir Arnold and Jürgen Moser showed that the feared perturbation of the planetary orbit in a many-bodied problem and the instability of the corresponding frequency relationships were sufficiently irrational. In other words, planetary orbits that are mutually harmonized in accordance with the model of the golden section indicate the greatest stability against disturbing influences in the heavens. Maximal irrationality makes maximal stability. It's no wonder that scientists as well as lay people call the resulting ratio and its visible manifestation beautiful.

Imaginals

The problem with a concept like irrational is that in everyday language the word gets used in derogatory and generally unpleasant contexts. This also applies to another technical term in mathematics—*imaginary.* This concept was introduced in science to define imaginary numbers that besides having ordinary (the so-called "real") components possess yet another dimension established by what is called the imaginary unit. On the advice of the mathematician Leonhard Euler, since 1777 this unit has been indicated with i (Fig. 3.11). The letter i stands symbolically for "root of minus 1," which is also written as $(-1)^{1/2}$, as you may remember from the classroom.

The Italian Girolamo Cardano introduced the root of a negative number in the sixteenth century. Cardano originally wanted to solve only the apparently simple equation $x^2 + 1 = 0$ and in the process endured the worst mental torture, as he tells it, in order to beg forgiveness of posterity. The equations in question can only be solved with the help of imaginary numbers ($x = i$ in the example), and to Cardano these new numbers didn't

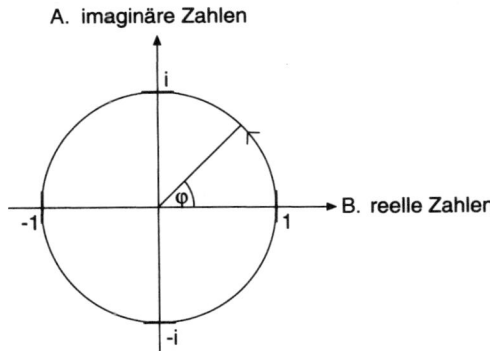

Figure 3.11.   The complex numerical plane with an imaginary unit in which a unit circle can be drawn, sometimes referred to in the field as "ring i." A. imaginary numbers; B. real numbers.

seem like God's creations. They were the work of human beings, and were so unreal they might possibly bestow the mathematicians with insights God had never intended.

Today scientists can't praise Cardano highly enough, for with a single stroke he put a symbol in their hands that offers them insight that would otherwise remain inaccessible. As a result, mathematicians today have known for a long time that the imaginary unit i is solely responsible for allowing them to understand division and other properties of prime numbers. However, they don't know why this is, and as a result some of them shy away from the beauty i gives its formulas, the aspect that Leibniz once called a monster between being and nonbeing.

More than anything else, it is surprising that modern physics cannot describe the world in mathematical terms without the imaginary unit. Whenever scientists debate the mechanics of the atom, they rarely fail to note the imaginary—the made-up—aspect of numbers. The imaginary is essential for a comprehensive presentation of reality, and in the meantime many physicists see i as the magic key that effortlessly reveals to them deep-seated symbolic structures of the material world.

According to Kepler and other scientists, the imaginary unit i makes elements accessible that belong to the mental world of images, awaiting their counterparts from the external world, reality. To better name the existential nature of this inner reality, the Frenchman Henry Corbin suggested in 1979 the concept "imaginal" to distance it from the all too pejoratively used expressions *imaginary* and *irrational*. The imaginal is not something that can shine in the light of day of rational scientific reason. It's part of the dark side of science and becomes visible mainly in the beauty of knowledge that succeeds in mathematical form. The beauty of an equation is then the sensory witness for something that we do not in fact understand but can perceive, through symbols.

. . . and Fractals

In the context of beauty and science, the geometrical figures called fractals and the books that portray the "beauty of fractals" are a natural subject.[9] There is no question that fractals can be beautiful (Fig. 3.12), but this alone does not make them art (Fig. 3.13). On the contrary! The attempt to shape fractal figures only emphasizes the difference between beauty in

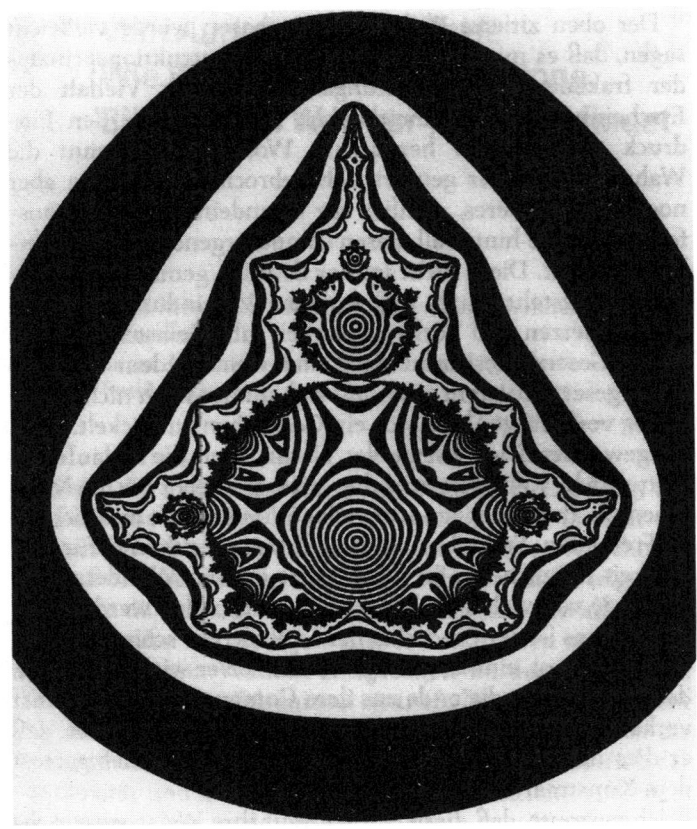

Figure 3.12.   A figure is called a fractal when it consists of lines that are not straight at any point. Characteristic for such lines or formations is the property of self-similarity, which we have also encountered in the appealing spirals (Fig. 3.9).

art and beauty in nature to which Kant was probably the first to call attention. I don't want to spend too much time discussing fractals but would like to raise the question of why the fragmented outline appeals to us aesthetically, why it seems beautiful. Most likely it is related to the fact that only natural objects have fractal dimensions and contours. The geometry of nature is fractal, on which one of the founders of this scientific direction, Benoît Mandelbrot, elaborated. In the same way that nature pleases us, in the multiple branch system of trees for example, fractals please us. They are beautiful without striving to be art.

Figure 3.13.   Cover picture for the magazine *Kunstforum* from December 1993 that promised "The new image of the world." Hopefully they don't mean the thing with the silly grin.

Hutcheson might say that knowledge of the structural principle of fractal images has led successfully to the discovery of uniformity in the variety of appearances, a condition that evokes the sense of beauty. Perhaps the perception of geometrically fragmented forms is the recognition of something more, perhaps behind all of that computer-generated beauty lies the fundamental insight of chaos research. This basic insight, which has not always been made clear, consists of the realization that there is no connection between natural laws and probability. Even when laws determine everything and we are aware of these natural laws, we still can't predict over a long period of time how an event will develop. In a positive sense, the laws that set the course of nature are flexible and this applies first and foremost to human beings. Regardless of science, men and women still have a sufficient supply of freedom. Fractals are the visible sign that freedom is possible. If we look at them this way, it's not surprising that many people see them as beautiful.

Mandelbrot, the man who discovered fractals, was once asked whether he believed the figures he pulled out of the computer could really be sold as art. His answer was that it wasn't his place to decide. He could safely leave it up to the fine arts market.

I suppose this answer is correct in a way, although only a small minority would find it satisfactory. Only in the fractals themselves can one find satisfaction, fractals discovered by mathematicians in their attempt to rise above the simple structures they thought they were observing in nature. The paradoxical situation is that they didn't get farther away from nature but further into it. Fractals dwell inside all natural objects. Fractals are beautiful—and may one day hang in museums.

4   *On Illumination of the Soul:*
How Pleasing Pictures Lead Science
to the Truth

Human beings long for intimacy with beauty
since in the face of beauty the wings of the soul grow.[1]

Paul Friedländer

In the spring of 1925 the young Werner Heisenberg, who worked as an assistant to Max Born at the Physics Institute of Göttingen University, was in so much misery from his hay fever that he had to ask for 2 weeks to travel to Helgoland to recover in the North Sea far away from blossoms and pollen. After being granted his wish, Heisenberg landed on the craggy island far away from the day-to-day operations of the university able not only to breathe in the pure sea air but able to struggle in peace with the physics problems that had driven many scientists in Göttingen and elsewhere to despair for decades—particularly the best among them. Physics as a whole seemed to be trapped in a dead end. Physicists felt, as Einstein once expressed it, "as if the ground had been pulled out from under one with no firm foundation to be seen anywhere." [2]

What they were searching for and hadn't found was the groundwork for a description of the movements of atoms, similar to the way someone might describe the motions of the planets and other celestial bodies. They were looking, in other words, for a new mechanics of the atom, and knew only that the old—classical—mechanics in spite of all their previous successes failed miserably when it came to explaining and making sense of the event at the atomic level.

Painful Recognition

The dilemma began shortly after the turn of the twentieth century when Max Planck failed to solve an apparently simple task. The task involved the colors that could be seen when solid materials were heated up and the light the material emitted when its energy rose. Planck had to learn from nature that it was not acting on a continuous basis and that one could find all the leaps that philosophers had refused and detested for centuries. Today we know that nature performs quantum jumps, but unfortunately the consequences of this insight have been too little understood outside of a narrow circle of physicists and mathematicians (and even today the concept of quantum leap is absurd in its trendy usage).*

After Planck's first step, the situation soon worsened because physicists had to admit that nothing fell into place any more in the world as soon as quantum theory made the rounds. Many scientists experienced it as a deep shock. From our perspective today, it's interesting to see how in the meantime there is no trace of this sensation left in science (in the scientific community). Without a personal inner experience, contemporary physics students are taught insights, which were disturbing in the past, purely as useful formulaic tools. They have to learn to use them and are not astonished by them. This is how today's students are presented with the beauty of physics, and as a result, even though it's right in front of their eyes, it remains concealed to them. Physics is practiced but no longer experienced. In fact, I have the impression that the modern form of physics instruction denies the beauty of science and renders it impossible for students to perceive it.

What are quanta anyway? When light and matter interact with each other, there is a fundamental, indivisible unit of energy that the light can gain or lose in exchange with the atomic building blocks of matter. This energy quantum can be calculated by multiplying the frequency of light with a very small constant number (Planck's constant) called the "elementary quantum of action" with *action* in physics defined simply as a product of energy and time (Fig. 4.1).

It all sounds so harmless in retrospect. For Planck and his colleagues, however, the world of classical physics had collapsed. We can try to

---

*Today's hottest new concept, "quantum leap," expresses that someone has taken a huge leap forward or will do so; this creates a dual problem: First, there is nothing smaller in nature than the quantum jump—besides zero—and second, by carrying out a quantum jump, atoms reach their fundamental state where they will ultimately remain. The quantum leap goes downward in nature and backward in slang.

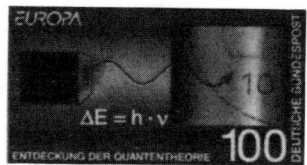

Figure 4.1. A commemorative German postage stamp from 1994 depicting quantum theory. In putting this new view of atomic reality forward, physicists were compelled to explain color. They succeeded when they recognized the energy (E) of light as a quantity proportional to the frequency. However, the price they had to pay for this piece of knowledge was very high (as discovered in the text). The stamp is beautiful for several reasons. The color is appealing, its spectrum is always attractive, and the stamp makes it clear that physics has something beautiful to explain. They try to do this with a simple system—a blackbody. The gap between black and the colors can only be closed when the quanta are introduced. Oddly enough, the quanta are of course themselves gaps.

imagine this catastrophe by observing a quantity that everyone at least knows by name—energy. The quantum attributes terrible things to it. On one hand, energy suddenly demonstrated its unstable nature, that is, appearing only as a multiple of the least (indivisible) unit. On the other hand, energy depended directly on frequency. Neither property could be reconciled with the heart of classical physics, the first law of thermodynamics, also known as the law of the conservation of energy.

Physicists and other scientists of the nineteenth century tried hard to prove that energy neither disappears nor emerges out of nothingness. They believed that energy, as indicated by many increasingly precise measurements, was only transformed, changing its shape between forms. Thermal energy, for example, could be transformed into motion, and kinetic energy could be lost in frictional heat, behaving similarly with electric, magnetic, potential, and other forms of energy. All of their many efforts culminated in 1847 in the original formula by Hermann von Helmholtz for the law of conservation, the first law of thermodynamics, which states that the energy of the universe is constant. Ever since then, this law has provided a solid foundation for classical physics.

The discoveries Planck was making threatened the idea of energy as a basic element of science that had to be conserved at all times under all conditions. How could a fundamentally discontinuous increase of energy exist at the same time as its naturally continuous (constant) conservation? And how is energy supposed to have (continuous) uniform duration per unit time if it is of a dimension proportional to the frequency of light? We can't even define the frequency of light for each point in time, only for time intervals. Do we eventually have to approach the thought that the law of conservation of energy no longer applies when it comes to atomic activity?

A Look at Beauty

Physicists did not immediately recognize the degree of change that they were proclaiming. However, between 1900 and 1925 it became clear that probably only a radical renewal of classical physics could incorporate the atomic facts and become its new mechanics. Many desperate attempts were made, even the occasional scrapping altogether of the law of conservation of energy, and many physicists could be found in a confused and excited state about the same time the young Heisenberg made his trip to Helgoland.

Heisenberg now decided to place his search for a new mechanics completely under the guiding star of the law of the conservation of energy and to hold on to it for all he was worth. In his autobiography *Der Teil und das Ganze* Heisenberg presents his initial premise at length, or to be more exact, he allows the reader to participate in his excitement without betraying what he actually did at that time, which was to develop a completely new formalism and to draft his own shorthand for it.[3] While in Helgoland, Heisenberg, without being fully aware of it, was giving the mathematical description of physical reality a completely new form. In other words, Heisenberg was a creative artist drafting a new image of the atomic world and then using it to start his work.

He was familiar with the model of the atom that Niels Bohr suggested about a dozen years earlier and that to this day determines the image of the atom with which most people are satisfied (Fig. 4.2). However, for Heisenberg, because the atom itself was considered invisible, this model relied too heavily on the visible or observable. Bohr's model, showing only the workings of a planetary system in miniature, was to Heisenberg philosophically completely unsatisfactory. The explanation of the (macroscopic) planetary system should not begin with a (microscopic) planetary system. Our internal assumptions determine our desired external results.

By drafting a completely new image of the innermost universe in order to get a secure foothold, Heisenberg clung to the law of the conservation of energy on which, as it turns out, he could rely. He described the decisive moment as follows:

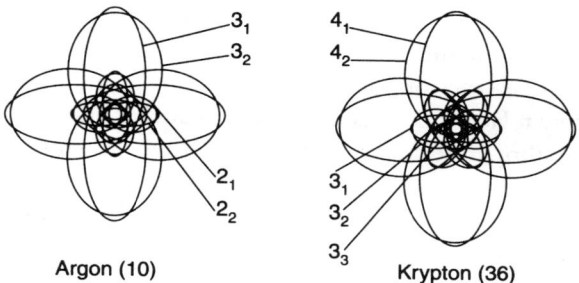

Figure 4.2. The observable Bohr model of the atom from 1912–13. The increasingly complex orbits are nevertheless definite paths traveled by tiny ball-shaped electrons that make a circuit around their corresponding atomic nuclei. Hydrogen is the most simple, with one electron circling around one proton. This is still how most people imagine the construction of the atom. It has been clear, however, for a long time already that there is no such thing as an orbit in reality, and instead we presuppose unobservable orbitals (technically speaking). We can't really make observable pictures of the atom, but images can still help us.

In this way I concentrated more in my work on the question of the validity of the law of conservation of energy, and one evening in Helgoland I was ready to go on to determine the particular terms in the energy table, or as we call it today, in the energy matrix, with a calculation that by today's standards was richly circumstantial. With the first energy state terms which actually confirmed the law of the conservation of energy, I became so excited that I kept making arithmetic mistakes in the subsequent calculations. It was already almost three in the morning by the time the final result of the calculation lay before me. The law of the conservation of energy had proved valid from top to bottom and—since all this came about on its own without being forced, I could no longer doubt the mathematical consistency and unity of quantum mechanics that it suggested. I felt as if I were looking from the clear surface of atomic appearances all the way to the depths at a foundation possessing a strange inner beauty, and the thought nearly made me faint that I was about to pursue the richness of this mathematical structure nature had spread out before me.[4]

Another Side of Science

A strange inner beauty was revealed with Heisenberg's final mathematical steps under the guidance of the law of the conservation of energy, and today we know that he was the first to have in mind what would later be taught to university students as quantum mechanics. The theory of the atom expressed by quantum is simultaneously one of the most beautiful and mysterious theories of science, and it seems to me that we can learn just how closely related both qualities—the beautiful and the mysterious—are if we are willing to roll up our sleeves and delve further into quantum mechanics.

The beauty Heisenberg grasped can best be explained by Plotinus, who lived and taught in the third century A.D. Plotinus confronted the Pythagorean definition of the harmony of parts with a definition that concerned the whole and defined beauty as the illumination of the eternal brilliance of the "One" through material appearance. Plotinus's thought is established in the Latin guiding principle "pulchritudo splendor veritatis," and with it, he takes science a critical step further. According to Heisenberg, the three words in quotes could be interpreted as "the scientist first recognizes truth by the brilliance [of beauty], by the light it gives off." [5]

Here is yet another reference to the illumination of the scientist's soul that Kepler had already known and explained and that Feynman knew as well. Heisenberg experienced it on Helgoland, and it helped him know immediately and unwaveringly that he was standing closer to the truth than any other physicist or other man of his time.

It's important to make clear what Heisenberg actually saw when he had his island experience. On paper before him were simply a few mathematical formulas that when later published (Fig. 4.3), probably did not release any feeling of beauty in anyone—whether the observer was educated in physics or was a newcomer in the world of formulas. The written signs and letters only become more than lines and figures on a piece of paper when the observer sees them as symbols. For a theoretical physicist like Heisenberg, there was always more to an "E" or a "n" than the abbreviations for energy or the frequency of light. In these signs, he could see symbols for a reality that was not concretely visible or otherwise comprehensible through the senses but was the reality containing the atoms. This second reality reveals itself through its perceptible influence. On this level, atoms are not things but are comprehensible as symbols, and the essential meaning of a symbol consists in the idea that it addresses not only thought, but also other human faculties, feeling, for example, and reaches the side of science that doesn't appear in the light of consciousness—and can therefore be considered part of the inner or dark side of physics.

This dark side probably also encompasses the One Plotinus discussed and of which Heisenberg was also convinced. This One could be summarized

Sei der Rotator repräsentiert durch ein Elektron, das im konstanten Abstand a um einen Kern kreist. Die „Bewegungsgleichungen" besagen dann klassisch wie quantentheoretisch nur, daß das Elektron im konstanten Abstand a eine ebene, gleichförmige Rotation um den Kern beschreibt mit der Winkelgeschwindigkeit ω. Die „Quantenbedingung" (16) ergibt nach (12):

$$h = \frac{d}{dn}(2\pi m a^2 \omega),$$

nach (16):

$$h = 2\pi m \{a^2 \omega (n+1, n) - a^2 \omega (n, n-1)\},$$

woraus in beiden Fällen folgt:

$$\omega(n, n-1) = \frac{h \cdot (n + \text{const})}{2\pi m a^2}.$$

Die Bedingung, daß im Normalzustand ($n_0 = 0$) die Strahlung verschwinden solle, führt zu der Formel:

$$\omega(n, n-1) = \frac{h \cdot n}{2\pi m a^2}. \tag{28}$$

Die Energie wird

$$W = \frac{m}{2} v^2$$

oder nach (7), (8)

$$W = \frac{m}{2} a^2 \cdot \frac{\omega^2(n, n-1) + \omega^2(n+1, n)}{2} = \frac{h^2}{8\pi^2 m a^2}(n^2 + n + \tfrac{1}{2}), \tag{29}$$

was wieder der Beziehung $\omega(n, n-1) = \frac{2\pi}{h}[W(n) - W(n-1)]$ genügt. Als Stütze für die von der bisher üblichen Theorie abweichenden Formeln (28) und (29) kann es angesehen werden, daß viele Bandenspektren (auch solche, bei denen die Existenz eines Elektronenimpulses unwahrscheinlich ist) nach Kratzer [*] Formeln vom Typus (28), (29) (die man bisher der klassisch-mechanischen Theorie zuliebe durch halbzahlige Quantelung zu erklären suchte) zu fordern scheinen.

Figure 4.3. An excerpt from Heisenberg's work from 1925 showing some of his basic formulas.

by Kepler's concept of the archetype and helps to explain how the formulas Heisenberg had before him on the paper allowed him to recognize the archetypal image from which his own thinking emerged. They allowed him to ascend the mathematical heights where he could finally see the law of nature—in its theoretical form—manifested by the same archetype.

Inside and Out

The influence of the archetype on a person's psyche and in nature may be a little overdone for a strict reader of the natural sciences. For someone versed in literature, the construction offers nothing new; we can read about it for example in Johann Wolfgang von Goethe's poem "Epirrhema," a term indicating that the lines serve as a small commentary.

> When observing nature you shall
> Always regard one as all:
> Nothing is within, nothing is without;
> For what is inside, that is out.
> So don't delay, seize it:
> Holy public secret.*

How much the internal and external can be interwoven is shown most clearly when scientists reach their findings—their realizations—in dreams or receive them in the form of inspiration. Psychophysicist Gustav Theodor Fechner, for example, was in a kind of a half sleep one October morning in 1850 when "in a slightly vague train of thought" it occurred to him to make a connection between "the *proportional* increase of the physical 'living-energy' of an object," which today we would call its kinetic energy, and "the measure of *increase* of its mental intensity." [6] Today this complex association, known as the Weber–Fechner law, builds one of the cornerstones of psychophysics.

More well known is the dream the chemist Friedrich August Kekulé had 12 years later when he was dozing off in front of a fireplace. At the time, Kekulé was trying to ascertain the structure of benzene (Fig. 4.4). Its annular ring structure, which today is a famous and substantial contribution to chemistry research, appeared to Kekulé first as an image in a dream:

> I turned the chair towards the fireplace and sank into a half sleep. Once again the atoms danced before my eyes. My inner eye distinguished larger images of multiple shapes, winding and turning like snakes. And then what did I see? One of the snakes took hold of its own tail, and the image swirled threateningly before my eyes. As if by a stroke of lightning I woke up and spent the rest of the night working on the consequences of the hypothesis. [7]

The Archetypal and the Alchemic

It's worth comparing both of these mathematical and chemical discoveries of a formula in detail. Of course both scientists only went public

---

*Müsset im Naturbetrachten/Immer eins wie alles achten:/Nichts ist drinnen, nichts ist draußen;/Denn was innen, das ist außen./So ergreifet ohne Säumnis/Heilig öffentlich Geheimnis.

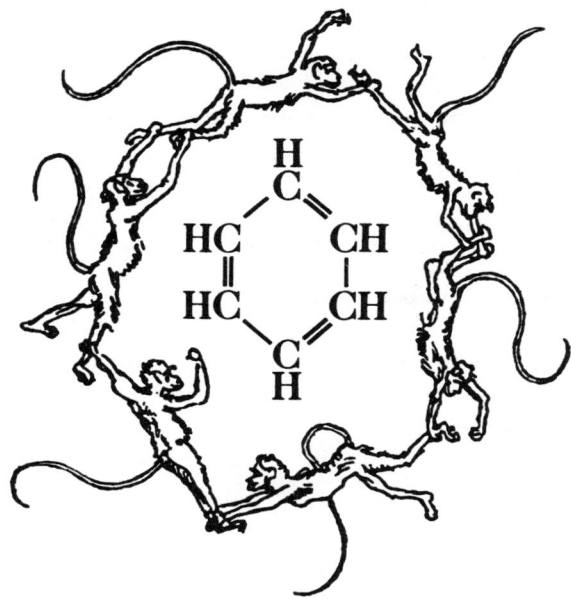

Figure 4.4.   The annular ring structure of the benzene molecule (in the center) with six carbon atoms (C) enclosed in a circle. Depicted is an illustration from 1886 (C. F. Finding: *Zur Konstitution des Benzols*) in which he explains: "in the same way carbon atom has four affinities, the members of the family of the four-handed have four hands with which they grip and hang onto other objects. Now just think of a group of six members of this family, Macacus Cynocephalus, for example, who form a ring with each other, where every other member uses two hands to grip: this is a perfect analogy to the Kukuléan Benzene Hexagon."

with their dreamed or envisioned insights after they tested the ideas "in an awakened state," as Kekulé referred to it much later. When he conceded that it was possible to find truth in dreams at all, he did so only on the condition that scientists do not allow their rational qualities to fall dormant. (This is self-evident and doesn't need to be continuously emphasized even if now and then the reader gets the impression that in the course of the present book this truth may have been forgotten.)

It's not hard to see, therefore, that intellectual work only begins after the dream or the vision of beauty. At any rate, getting a glimpse of the truth sends scientists into a state of excitement that can easily make them stay up the whole night to work. Like addicts, they can no longer control their urge. They are compelled to search for the solution, for the truth. This is an observation that isn't likely to be found in conventional depictions of science and its history, although it has been more than represented in literature. In the novel *Der Mann ohne Eigenschaften*, Robert Musil's protagonist, a mathematician with a sense of possibility, says

> Knowledge is a behavior, a passion. Essentially a forbidden behavior; for like an individual addicted to alcohol, sex, or violence, an unbalanced character develops the urgent need to know. It's not accurate to say that a scientist hunts down the truth; it hunts him down. It makes him suffer.[8]

Musil addressed another aspect of the dark side of science—instinctual drives. As important as this aspect is, however, the act of recognition and the parallel between Heisenberg's island experience and Kekulé's fireplace dream is more relevant to our context. Both relied on the archetypal and on the ensuing archetypal images or ideas that helped them succeed. With Kekulé, a snake is biting its own tail, and it seems to me that this image has not emerged by coincidence. After all, there is no older symbol that depicts how one comes from all and how all flows in one than this body enclosed in a ring, which holds an important place in the history of alchemy (Fig. 4.5). The methods of modern chemistry may have little to do with the processes of the early alchemists, but anyone moderately

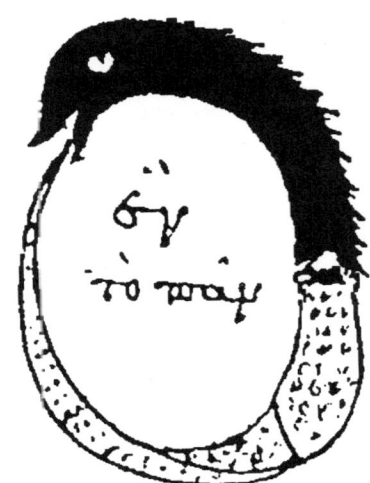

Figure 4.5. An Ouroboros biting its own tail. The writing inside the body can be translated as "The One, the All"; from the *Chrysopeia of Cleopatra* Manuscript 2325 of the Biblioteca Marciana in Venice.

familiar with the history of science and who takes its goals and its thought processes into consideration knows that alchemy may have vanished from laboratories, but not from our minds.[9]

Officially science historians tell the public that the alchemical–magical form of describing nature in the seventeenth century has been discarded and replaced by the experimental–quantitative form since practiced in science. However, in the eighteenth century, Newton was still strongly drawn toward the alchemical notion of unity, and its allure led him to understand that the force exerted on earth below (which allows an apple to drop out of a tree, or to be more precise, pulls it down) is the same force that keeps the moon in orbit high above in the heavens. Newton also believed in the possibility of the full transformation ("One from All and All from One"). In his day a laboratory was still called a *laboratorium*, with echoes of the word *oratorium*. This associated perception has long since faded from the operations of today's science.

Newton's alchemy is by no means a unique case, as is shown by an area of science geographically and temporally distant. Once again I am referring to Goethe. In the second part of his altogether alchemically concerned *Faust*, the homunculus is produced in a laboratorium where in terms of method, everything is conducted alchemically; human matter, for example, is to be enclosed "in a retort" for the creators to "cohobate" or distill repeatedly. Additionally, anyone who takes the idea of the big bang seriously should realize the extent to which it represents alchemical thought; the point in time designated for the event can only have been the "One" that arose from the universe or the "All." In the alchemical–aesthetic moment of the big bang we can only imagine One thing that contains the elements that later make up All (the universe) space, time, energy, and matter (Fig. 4.6), something like the modern version of four classical elements: fire, earth, water, and air.

I wouldn't want to develop a teaching method based on the history of alchemy but I do think that we can use it to understand the fundamental thoughts of the European scientists when they attempted to understand the world. This even includes archetypal material like the circle or the ring, which scientists in all ages have searched for and found—either as planets that move in an orbit in the sky, as the electron orbital in the atom, as the cycle in biochemical metabolism (Fig. 4.7), or as the circulation of blood in the body.

The circle was to Kekulé what energy was to Heisenberg, who was dreaming if not sleeping. Heisenberg's dream was to fulfill the conditions

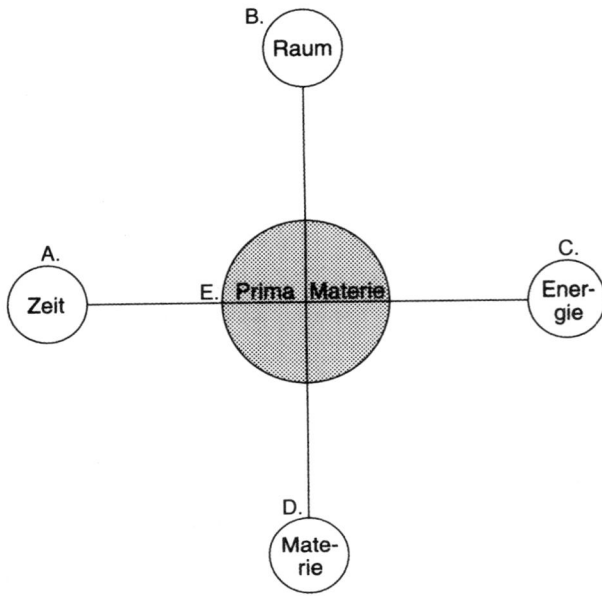

Figure 4.6.   The alchemy of the big bang is only made possible (thinkable) by a
thought from the theory of relativity. As Einstein himself was always trying to
make clear, the essence of his theory of relativity is that it is no longer possible to
allow things (their mass and energy) to disappear and leave space and time
emptied. When things disappear, space and time disappear as well. Those who
believe the big bang was possible and consider themselves in the right scientifically
think like the old alchemists, who were certain that the elements of the world could
be traced back to a *prima materia*. I mean not to accuse with this reference to
alchemy but to praise. Alchemy was a beautiful science, and it produced beautiful
images and beautiful thoughts. The reason for its failure is anything but trivial, and
it deserves to be pursued more thoroughly. In my opinion, the answer has something to do with the fact that alchemists sought unity as a concrete visible reality.
Today, scientists know that they can only strike it rich on another level. A. Time;
B. Space; C. Energy; D. Matter; E. *Prima materia*.

of the law of conservation of energy. Energy offered him vision and
purpose, and this is remarkable because we cannot define with any
amount of precision what energy actually is. Physicists may mumble
something about an abstract quantity of an unchanging system, but without knowing it, the famous British astrophysicist Sir Arthur Eddington at
the beginning of this century was closer to the truth when he used the

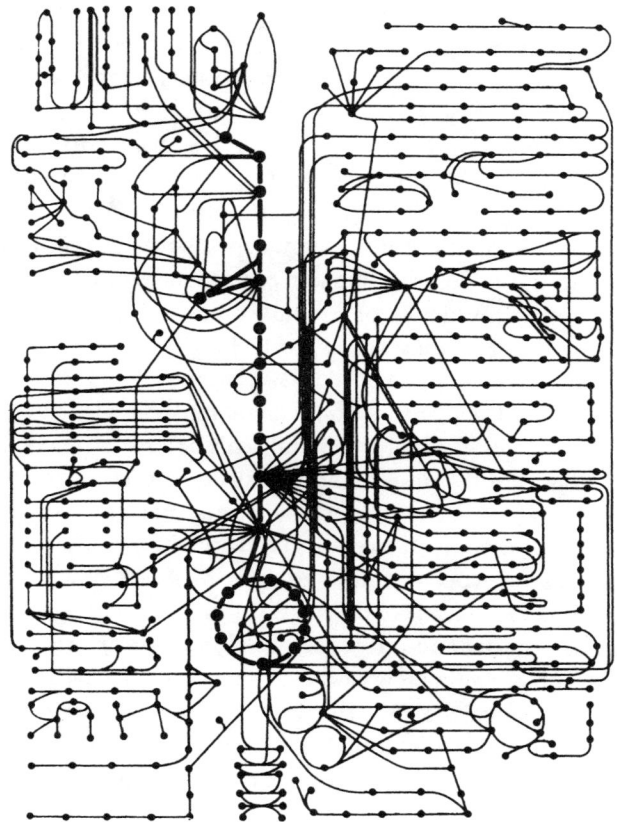

Figure 4.7. Schematic presentation of part of the metabolism of a liver cell. Especially prominent is the breakdown of the sugar glucose that takes place in the citrate cycle (from "Mannheimer Forum 89/90"; p. 150).

expression "energy of something."[10] A similar idea is discussed in a new textbook on energy by Klaus Heinloth, which begins with the sentence, "We can use energy in many ways; we can almost perfectly describe energy even in its various modes of appearance; but in the end we do not know what the phenomenon energy is."

If we look at it this way, to consider science not in empirical but in archetypal terms is no longer a very surprising concept. The point is that it presents a fundamental idea in the position to take on many forms, internally as well as externally, which human thinking cannot do without.

In any case, energy itself possesses a considerable alchemical quality—*transmutability*. In every kind of event, energy is converted from one form into another, and we run into problems in modern science when we don't understand how this process takes place concretely.*

When Heisenberg and other physicists are drawn to the law of the conservation of energy it's not because it's been proven, but because they know from their innermost conviction that nothing can be lost, only transformed. Energy is something like a canvas for souls to paint the pictures that precede thought. For this reason physicists would not accept any experimental counterevidence of the law of conservation of energy and they searched as long as they could for technical errors until they found them.

Heisenberg's search for the atomic laws makes it clear in my opinion that archetypal images guide the rational search of researchers. Even more interesting is that it also brings to light that the end of all of our efforts—the atom—is itself of archetypal dimensions. For how else are we to understand how the concept has survived for thousands of years while what we mean by it has been completely transformed? When the atom—in Greek, the indivisible—was conceived of in the classical age, it resembled something concrete that could be assembled like building blocks to create the world. This observable view of the atomic plane has become one of the distant past ever since the first decade of the twentieth century, when quantum mechanics was developed. The visible world was replaced by an atomic reality that could only be represented symbolically.

Decisions

The new mechanics of the atom, which had its first breakthrough with Heisenberg's beautiful view, made not only friends among physicists, it had its enemies as well. Probably the best and certainly the most famous of them—Einstein—spoke out repeatedly against this theory of the atom to the end of his life, although it is important to keep in mind that Einstein didn't believe that the theories of quantum mechanics were false. He just

---

*For example, take the question of the nature of the mind. We run into difficulties as soon as we separate it from matter. When the mind is not produced from matter—if it is complex enough—but when mind is identified *as* matter, in the form of the brain, the question is raised of how it transforms itself into the energy necessary for the chemicals in the human head to keep thought going.

refused to see them as the final word on the subject; Heisenberg's and others' descriptions of the atom seemed incomplete to him. Einstein refused, moreover, to grade it with even a "B" (or value it as beautiful) because it didn't fulfill his requirement that a physics theory leave no room for the random. "God doesn't play dice," according to his famous dictum. Enough has certainly been said already about the intellectual shortcomings of this commentary, but nothing has been able to detract from its influence. Einstein used the right form—a biting aphorism—in response to his critics, and this aesthetic charm helped lift his words above and beyond any inherent error or weakness.

There will be further opportunity to discuss this aspect of aesthetic appreciation and denial of quantum theory. I would like to turn now to other contributions Einstein made to physics that likewise can't be understood without recourse to beauty, this time even with positive signs, although as yet it hasn't been discussed much among biographers.

The French mathematician Jacques Hadamard had already referred in the mid-1940s to aesthetic aspects in the works of great physicists when he ventured to discover the psychological aspects of theoretical research and rational cognition. Hadamard formulated the thesis, which became a certainty to him, that for every scientist there is a critical moment in thought when he or she must choose to follow a course according to aesthetic principles and to decide in favor of beauty. The choice is usually between different sets of assumptions for the basis of a theory that cannot be settled by an experiment.[11]

To mention one example, around 1912 Niels Bohr had to decide between classical physics and a model of the atom as a planetary system in miniature. Experiments in which a ray was produced from the nucleus of an atom and scattered on a thin gold foil suggested an image of the atom in which electrons revolved around a nucleus the way planets do around the sun. Because electrons, as opposed to planets, were charged, according to the laws of classical physics they could not move in orbit without giving off radiation and losing energy. In other words, the beautiful and transparent image of the atom didn't fit with classical physics, and Bohr had to choose one or the other.[12] He decided in favor of the image and resolved to transform physics. His choice was thus highly irrational. It can only be understood aesthetically, as Hadamard maintained, and it set one of the most significant scientific upheavals in motion.

Hadamard, however, had Einstein's path to the theory of relativity in mind when he formulated his hypothesis on the choice for beauty. He could do so because he had recourse to corresponding conversations with Einstein. Einstein himself had continuously emphasized the psychological—spiritual—components of scientific research and in response to psychologists spoke about many of the images preceding his thinking. It was difficult for him most of all to convey his thoughts to others on things he had long understood only visually, first in formulas and then in words. The decisions Einstein made were related to the desire for beauty.

The deciding condition for the first form of the theory of relativity, today called special relativity, is accepting that the speed of light is constant. When Einstein chose in favor of this premise in 1905, no corresponding experimental evidence existed. Einstein himself knew best that the constancy of the speed of light at first glance appeared to contradict a further condition of the theory. This second hypothesis is called the Galilean invariance, and it states that all systems that include motion are physically equal when they fulfill *one* condition. No forces in the system can have an effect on a uniformly moving body. Experts talk about inertial coordinate systems and we can imagine two such systems in the form of a train station and a moving train. Physical laws have the same form for the train that does not speed up and thus moves at a uniform rate of speed, as they do for the train on the platform. The Galilean principle demands this. In other words, a passenger on a train and a passenger waiting on the platform at the train station experience the influence of the same forces.

It is not a question of how Einstein or his theories established that for both the traveling passenger and the waiting passenger the speed of light is invariant, meaning the way he took into account the today proven fact that light leaves a flashlight that is moving in the train just as fast as a flashlight moving on the train platform.* I am far more concerned here with the psychological significance of Einstein's decision to assume a constant speed of light, and why this was so impenetrable for his colleagues.

Ether Is Dead—Long Live Ether

What Einstein did when he assumed an unchanging speed of light was to do away with ether; physicists still needed ether to provide a

---

*Technically it undergoes the Lorentz transformation that replaced the old Galilean transformation without losing the required Galilean invariance.

medium of light for the universe. They knew that light could spread through the entire cosmos, and since it passed through a wavelength, there had to be some medium for light waves in the universe. Liquid waves needed water, sound waves needed air, light waves needed ether. Just as water and air, seen abstractly, present a kind of mechanical state of stress, so did ether. The question that preoccupied many physicists in the second half of the nineteenth century was as follows: "How can the rotation of the earth be conveyed relative to this ether?"

Einstein had enormous difficulty in revealing this subject as a pseudoproblem. The difficulty had very little to do with the intellectual. If this were the case, physical expertise would have done away with ether a long time ago. It was a substance that would have had to be harder than steel and thinner than air to be able to move light—something that popular wisdom would have recognized as an impossibility in no time.

The difficulty physicists had in dispensing with their conception of ether's role in the universe had more to do with psychology than with physics. Western scientists since the classical age, beginning with Aristotle, have spoken of ether. Its enduring popularity lends all the more credence to the proposition that we are dealing with an archetype that we cannot distance from Western thought without entirely stifling it. Einstein had let out the "physical air" that filled the cosmos and left the universe behind empty, like a flat tire.

Or only at first glance was this the case. At second glance we see that Einstein quickly filled up the space with a new "ether," easy to understand if we assume that it represents something archetypal. In current physics, Einstein's new ether is considered a field, as in gravitational field, and the essential difference is in the epithet "mechanical." Whereas they imagined the old ether as a mechanical state of stress filling space—as even Faraday conceived of the first fields (Fig. 4.8)—Einstein's field turned into a new, second form of reality, no longer directly observable but provable through concrete influences.

In 1905, the young Einstein did not have all these implications in mind when he assumed a constant speed of light. What interested him immediately at that time was the possibility of working into Newtonian mechanics what the Maxwell electrodynamics already possessed—an invariable rate of speed. Einstein sought uniformity among variety—a beautiful theory in the sense of Hutcheson. I am convinced that Einstein had nothing else in mind; and Paul Dirac, the English physicist and Nobel prize winner, expressed this clearly when he wrote that Einstein when

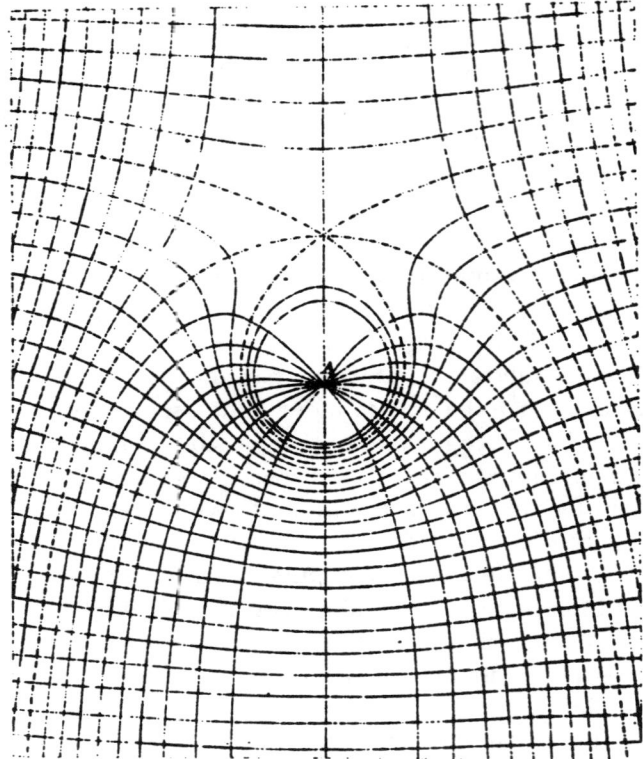

Figure 4.8. The lines of a magnetic field that Faraday postulated and made observable in a (beautiful) drawing from Maxwell's *Treatise on Electricity and Magnetism*. Here we can see a magnetic field generated by a uniform field and a wire with a current flowing through it.

working on constructing his theory of gravity, didn't try to take into consideration any kind of observational results. Dirac maintained that nothing could be further from the truth and that Einstein's process consisted exclusively in looking for a beautiful theory.[13]

Dirac even expressed the view that Einstein's theories still interest us today because of their beauty and would fascinate us even if not supported down to the last experimental detail and—seen pedantically— were false. For Dirac, it was important that a theory pleased on account of its beauty. Consistency with reality was just a nice addition. Dirac viewed the advancement of a theory as the creation of a work of art. The New

Zealander Ernest Rutherford, who as a physicist was distinguished with the Nobel prize in chemistry in 1908, expressed this similarly: "Einstein's theory of relativity can only be seen as a glorious work of art, completely apart from questions as to its validity." [14]

Einstein's Four Elements

To be utterly accurate we would have to say that Dirac's comment was referring to Einstein's general theory of relativity. This theory distinguishes itself from the already mentioned special variant in that not only observers moving in a uniform rectilinear motion with respect to each other are to experience the laws of physics equally, but all observers are. Behind this apparently absurd claim is the statement that on the earth we are in anything but an inertial coordinate system without force and despite this we are able to formulate the laws of physics. Because our location is arbitrary, every other observer must be able to see things the same way. In other words, the physics related to gravity and the universe must be formulated in such a way that its laws are independent from every possible location of an observer.

When Einstein set out to work on this problem it was considered unresolvable; in fact, for more than 10 years he sought the correct description whose essential quality today is characterized by the concept of covariance (which covers far more ground than the Galilean invariance or special relativity). The end product of Einstein's effort is, in Dirac's view, a theory of great simplicity and elegance whereby two aesthetic key words are conspicuous, and worth illustrating.

Einstein sought a simple initial condition, as in special relativity, on which to base his thinking. In his first great draft of 1905, what was formerly the constancy of the speed of light became the identity of inert and heavy masses in the general version, which in any case is simple to formulate. It is also easy to understand if we understand mass as a property of objects that resides in them independent of external circumstances. Through their mass, objects become heavy, meaning they receive weight, and through their mass, objects become inert, meaning they resist acceleration. Mass becomes heavy through a force—gravity. Additionally, mass becomes inert against a force, that is, one used to accelerate it.

If we look at it this way, it is impossible to bring both masses together, but one day Einstein had the "happiest thought of [his] life." [15] He was startled by his sudden vision that someone falling from the roof of a house would no longer feel his own weight, or gravity. The acceleration that the person would feel during the fall neutralizes the gravity. We can express this by saying that gravity acts like an accelerant, and heavy and inert masses are equivalent. Einstein's thought also began with an image, and an image that is convincing and fascinating in its simplicity. Today we are familiar with this image and have been for a long time, from television, for one, where we've seen astronauts turn around and around inside their gravityless capsules while they swim behind their toothbrushes.

The actual difficulty of bringing the initial conception into focus mathematically as an exact science was that at some point the equations became more and more complicated and the coveted covariance seemed to develop into a nightmare. Einstein set this matter to rights by associating gravity with the geometry of space, where the force of gravity developed. To make this connection, he had to make the geometry of space independent of the mass that existed in it, creating for himself further mental grief. He introduced something called the curvature of space, implying a deviation from the axioms of Euclidean geometry, and he made gravity dependent on it (Fig. 4.9). Under these conditions the equation could become

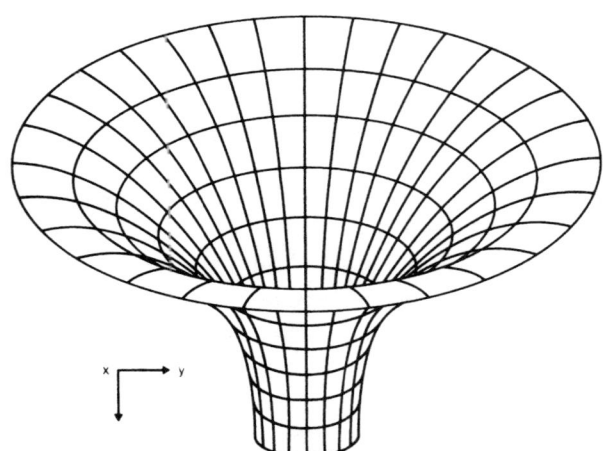

Figure 4.9. Einstein saw space, time, matter, and energy as interconnected, not isolated. The most interesting notion is that the geometry of the cosmic space is dependent on whether it bends around mass. Mass bends space, and the denser the space is, the more forcefully mass bends it. In a borderline case of very high mass density ("black hole"), (again cosmic) space then looks like the depicted funnel.

covariant, and Einstein was able to draft a theory of the universe—a cosmology.

The theories that Einstein brought into existence after his aesthetically motivated decision boil down to the recognition that elements that we see as separate and independent in truth belong together. Mass bends space; space is woven together with time; time is dependent on energy; energy is the equivalent of mass. In other words, Einstein created a connection, a *uniformity* in the variety, in a way that could not be more profound or unified. His theories are in fact works of art, just as Rutherford expressed.

Einstein's vision of the world has introduced a fully new view of the fundamental physical dimensions. Before, we had always imagined space as independent of other conditions; when things disappeared from the world they left behind an empty space. Einstein demonstrated in his portrait of relativity that even space departs along with the things—their mass and their energy—and so disappears. All becomes one. And in reverse it means that all must have been created from one. This is a reference to the big bang theory, and anyone who is curious about why this idea keeps such a tenacious hold on our imaginations despite experimental difficulty and much opposing astronomical evidence should consider what is said about it in the illustration in Fig. 4.6.

Indeed, there is overwhelming evidence that Einstein wasn't too amused by the idea of the big bang, but this didn't change the fact that his vision of the world clung to something archetypal, because there are still, as in the classical age, four elements. Instead of fire, earth, water, and air we now have space, time, matter, and energy. Additionally, whereas Aristotle postulated a *prima materia*, an original material from which the archaic group of four could originate and become influential, Einstein went on the hunt for a unified field theory which took on the same task of the *prima materia*.

That Einstein never found his unified field theory is no objection to the intellectual ideal his thought represented. In the end he believed in a unified field as much as Aristotle had in the original material. They both decided in favor of a uniform primary component because the thought was and is beautiful. It is beautiful because the thought arranges the variety in a way that makes the unit recognizable. That was true with Aristotle; it was true with Einstein; and it will continue to be true for a long time to come.

5   *Beauty in Revolution: The Aesthetics of Scientific Progress*

The true subject of science is the beauty of the world.[1]
Simone Weil

Sometimes science makes philosophers clueless. Why is it, they keep asking, that research in the natural sciences is so successful and that very often it is actually possible to know and explain a thing? Not only the blue of the sky on a cloudless day or the red of clouds at sunset, both brought about by the simple scattering process of particles, but also all of the colors of a metal heated long enough to show its incandescence. To be able to explain these colorful pallets that depend on temperature, physicists have to know not only that there are atoms, they also have to understand how these atoms are built and what gives them stability. How do scientists know this? How do they know that they know anything—for example, that planets move in an elliptical orbit around the sun, or that atoms have a positively charged nucleus and that negatively charged electrons move around the nucleus symmetrically but do not make a classical orbit? How do they know that light travels across a vacuum at the speed of light in the form of electromagnetic waves and can at the same time appear as a packet of energy? How do they know that bacteria have genes whose variations (mutations) take place spontaneously and arbitrarily and are not induced? What is so special about scientific access to the world that brings more certainty in knowledge than all other attempts at thought (although it can go wrong of course, tremendously wrong) and can at its leisure even take aim at infinity, whereas those searching purely philosophically threaten to be stuck in an apparently terminally tangled undergrowth of paradoxical concepts?[2]

No Logic in Scientific Discovery

Friends of scientific discovery like to point out that science is different from philosophy because under restricted conditions scientists seek explanations for natural phenomena—in other words, we experiment and question nature. The anecdote generally told at this point is of the Danish physicist Niels Bohr who was on a ski trip with some friends, including Werner Heisenberg and Carl Friedrich von Weizsäcker. It was evening in the cabin and Bohr was on dish duty. After he had started clattering around in the kitchen and had immersed the dishes in water, all at once it grew quiet. The great Dane had fallen into contemplation; suddenly they heard him yell:

> Now I know how science works! It's like washing dishes. Here in the dish tub I have dirty dishes; they're swimming in the dirty water and will be man-handled with a dirty rag. In spite of this, the dishes and cups will be clean afterwards. In science I begin with muddled concepts, I test them in muddled experiments, and I interpret their results in muddled language and opaque grammar. In spite of all this, I am certain to know something at the end of this exercise that I did not know before.[3]

Bohr was said to have gone on with his rant to say that philosophizing was like washing dishes without water, because a philosopher wipes muddled concepts with muddled logic, but this is another story for another discussion. The washing image expresses the basic idea that the interplay of the three elements is instrumental to the scientific experiment, and that the scientific experiment itself is designed to test the durability of "muddled concepts" in nature.

A scientist might have maintained at one time, for example, that on the far side of the moon there was a blue unicorn dancing the tango. However, no one would have taken this seriously as a hypothesis until there was a way to test it in an experiment, say in a moon flight. After the actual moon flight in the 1960s, during the Apollo project when no one observed a colorful fabulous creature performing any kind of activity on the moon, the hypothesis would have had to be taken back. It would have been exposed as false.

The unicorn example may be far-fetched, but it still represents more or less how many scientific theorists and practitioners today continue to regard the progress of scientific research that Karl Popper most distinctly introduced in his famous book *Logik der Forschung*, published for the first time in 1934.[4] Popper, who is related only by name and not by blood to

Josef Popper, defined the essence of science as setting up and testing theories. To be more precise, individual scientists set up hypotheses and test them in experiments. Popper's decisive contribution consisted of theorems with scientifically provable hypotheses that in an experiment— an observation or an investigation—*could be proven false*.* The hypothesis, according to the technical expression, must be falsifiable, and in the logic of scientific discovery, an experiment where a hypothesis is falsified furthers the ongoing search for knowledge; a new hypothesis must be proposed, a hypothesis that Popper (understandably but groundlessly) would consider better. The reverse situation, a corroboration through experiment, allegedly does not exist. The verification of a statement in an experiment may be very nice, and it can also lead to a doctoral degree, but according to Popper, this kind of result doesn't help further science as a whole because the researcher only sticks with the previous formulation. No rethinking takes place and knowledge doesn't change. The corroboration doesn't produce a new piece of knowledge. Only falsification leads to progress and, as Popper assured, it will then be steady and ascending. And how else? The following lines that were quoted widely in my student days might help us to better understand the scientific theorist camp and why scientists always know their way around better when their method is scientific.

> The way to wisdom, I explain,
> Is easy to express,
> To err and err and err again
> But less and less and less.

The quintessence of inductive logic expressed by this quatrain is that we are never sure of knowing the truth and we can only, from one attempt to the next, minimize our errors.

The concept goes back to the great Francis Bacon who, in the seventeenth century, first brought attention to the problem confronting experimental scientists who were attempting to draw general conclusions from single observations. If they saw a white swan, they would draw the conclusion that all swans were white (which was falsified after black swans were seen). Or on observing that peas inherit their properties according to fixed hereditary principles, they would draw the conclusion that all living creatures are bound to this model.

---

*That the statement about the blue unicorn dancing on the moon falls into this category might be guessed.

Bacon mentioned, 300 years before Popper, that a single opposing example was enough to bring down a hypothesis, and that research only arrives at something new when a single falsifying experiment forces the scientist to think of another hypothesis.

Thus, a hypothesis belongs to the realm of empirical science when it is in principle falsifiable, and research makes progress when this happens in practice; thus a hypothesis in the experiment is proven false when a black swan is sighted, a unicornless moon is observed, or an organism is discovered that follows no known hereditary principles.

Many philosophers imagine that science operates this way, and many scientists do too. In the end it is nice to work on philosophically blessed terrain, and yet neither philosophers nor scientists see that everyday research and the business of science has nothing at all to do with this simple model. No one abandons a hypothesis and then an entire theory because some experiment went wrong. In the first place, most of the technical apparatus used is far too complicated, and in the second place, each result must be interpreted, an element the "logic of scientific discovery" does not take into account. When I investigate the hypothesis that bacteria not only have genes, but that these genes can change through recombination, I have to be fully aware that in an experiment we can only count bacteria and its molecules, analyze them, or do both. The gene in question is only an idea to help me understand the living material. I have to translate every experimental result into my ideas, and inductive logic and falsifiability are irrelevant at this juncture. Here the scientist is alone with his or her interpretation.

"Beautiful—Publish"

At best, we can say that the value of Popper's theory of scientific progress consists in its capacity to raise the ambitions of the scientific method, and it has this capacity precisely because it is easily falsifiable. To falsify Popper's theory, we would have to make a historical science experiment, meaning we could take a look back into the history of physics, for example. We would have to note that there was no experimental proof for Copernicus and his system for centuries and then when it did come, it came in the form of a verification (!) by Bessel (Fig. 5.1). The same holds true for the basic premise of Einstein's theory of relativity that was only taken seriously for the first time in the 1960s and not proven false, but confirmed.

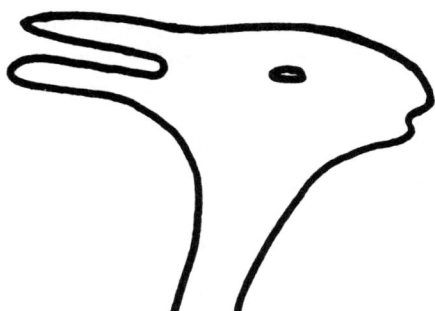

Figure 5.1.   The psychologist Jastrow's ambiguous duck–rabbit figure. Either the observer sees a duck with its beak facing left, or a rabbit with the same lines representing ears. In both cases, the dot marks the eye. See also Fig. 5.2.

Einstein is a perfect example of how little logic goes into scientific research. In 1911, the news reached him that in French laboratories under the leadership of the later Nobel prize-winning Jean Baptiste Perrin, one of his theories (of diffusion) was tested and as a result deviations were said to have been found. Completely unknown in those days, Einstein responded by writing the widely known professor from Paris with the news that he had already recalculated everything and after an insignificant correction "there was no other error," unmistakably implying that the fault lay squarely "on the experimental side."[5] In other words, the first thing scientists look for is a technical error (in the Einstein–Perrin case, they actually found one) before they waste time thinking about giving up an appealing theory, especially, as in Einstein's case, one that "fully satisfied" him on account of its beauty.

Especially interesting in this regard is the oil drop experiment that the American physicist Robert Millikan used to determine the magnitude of the elementary charge. Millikan's hypothesis was that electric charges in nature are not distributed continuously but only in discrete units, as whole number multiples of the elementary charge that something like an electron might carry. Millikan wanted to determine the magnitude of this charge with an oil drop charged with electricity that was held between capacitor plates in suspension, with the capacitor set up so it counteracted the gravity that supported the drops.

Millikan received the Nobel prize for physics for this experiment and his results, and even if we overlook the fact that nothing in his experiment could be falsified, what is more significant in our context is what the science historians discovered after Millikan's death, when they took a look into his lab notebooks (Fig. 1.8). There in black and white is stated that Millikan had only published the experiments that corresponded to his hypotheses and that seemed accordingly *beautiful* to him. He looked above all for the "beauty" that he had had in his sights from the outset.

"Anything Goes"

These and other findings made in the 1950s and 1960s show clearly that scientific theories in no way are derived from compelling logical conclusions from cleanly conducted books of protocol that are subsequently tested in painstakingly conducted experiments. This version is a philosophical fairy tale. Scientific knowledge is produced by thoughts based on empirically discovered and experimentally tested factual findings, but no one said anything about logic. Other factors share the main role, and I have already discussed several of these. How is logic of all things supposed to explain how a scientist arrives at one hypothesis or another? How are we to explain how hypotheses come about at all? A scientist's brainstorm can be almost anything—surprising, fantastic, or crazy, for example—but it is hardly logical.

When at some point the scientific philosophers began to realize what science historians had known all along, some of them couldn't keep it to themselves and immediately announced new theories about the progress of scientific knowledge. They were clinging to a thought fundamental to Popper's logic, the basic thought that science can only become better and its insights clearer and more profound: Einstein knew more than Newton who knew more than Galileo who knew more than Copernicus who knew more than Ptolemy who knew more than Aristotle and so on, although we cannot therefore conclude that Aristotle knew nothing. In fact, the opposite is true; Aristotle already understood everything, something that is incompatible with the logic of scientific discovery.

The new scientific theory became popular through the catch phrase "anything goes" that spread the thesis in an especially easy and accessible way, implying that science requires no commitment to method and that it is not worth the trouble to look for the systematic ways scientists have used in the past in their search for the truth. In the long term, more important than this methodological anarchy is the statement that two kinds of science are to be differentiated for a successful theory. On one hand, we have the supposed operation of normal science where many measurements are made and data are generated in the scope of a given frame of thought. On the other hand, we have upheaval, which subjects the old order to increasing amounts of incomprehensible information so it collapses under its weight, and after a chaotic phase, a new order is achieved.

Most of the time—according to Thomas Kuhn's thesis in his famous book, *The Structure of Scientific Revolutions* written at the beginning of the 1960s—scientific researchers in conventional science concern themselves with the solution of riddles to magnify their experiments, accuracy, and degree of factual knowledge (for example, the boiling point of compounds or the compressibility of matter), taking up a considerable portion of the literature of experimental and observable science, and it is interesting to note that far more than half of the work by conventional researchers is never quoted by anyone not to mention the authors themselves.[7]

Most of the work is done by those souls undertaking research who are trying to make their small contribution toward the great adventure called science. For them, research is central and can even be exciting. For others, however, normal science seems extremely boring and as a result loses its appeal. The activity of this normal science, at least according to outward appearances, produces neither realization nor discovery. Such a researcher simply collects more and more information and piles it up until, at some point, it can't go on any further and the many pieces of the puzzle produced and delivered by research centers at universities and institutes no longer fit together. At first, the pieces of the puzzle are scattered around and don't even come close to fitting together, but as soon as someone perceives the monumental mess they have created, science becomes interesting.

What takes place when suddenly a new way of thinking or a new concept of order is not only necessary but also delivered has nothing to do with logic. Kuhn introduced the concept of a *scientific revolution* in order to give a name to the change of views that apparently occur under pressure of empirical data. The passage from classical physics to quantum physics or the emergence of the theory of relativity are usually quoted as examples of scientific revolution. Also considered revolutionary are the emergence of evolutionary thought in biology in the nineteenth century and the successful founding of chemistry in the preceding century when oxygen and hydrogen gases were discovered in the air.

At any rate, the word *revolution* describes a circular flow in which we arrive again where we once were. Revolution refers thus to a closed cycle that can be symbolized by an Ouroboros (compare with Fig. 4.5).

The concept of a scientific revolution or revolutionary science is considered somewhat debatable or problematic and is referred to in less than praiseworthy terms. This is where the circle, the cycle, and the Ouroboros come in. Even though we can't say for sure what a revolution in science is—what about the discovery of molecular biology?—we hold on steadily to the idea. It seems to me that the popularity and persistence of the idea of revolution can be attributed to its geometric form—the circle. When I use the concept of revolution, I am referring to my vision of the external world through an archetypal image existing deep within me and that I find appealing. We generally like to let these images finally rise to the surface and use them in relation to the external world.

Perception in Science

This last concrete example, oxygen in chemistry, might serve best to illustrate a concept that Kuhn considered integral to describing scientific revolutions. Kuhn believed that all the practicing scientists of a generation make silent assumptions about the world and the nature of things, and under these conditions they interpret their results. He compared the paradigm of science to a conceptual box firmly implanted in the minds of all of the researchers of any given generation.

The paradigm of the chemist has been that besides earth, water, and fire, air is one of the four elements and as such cannot be broken down any further. The concept of gases emerged, however, in the seventeenth century and soon included steam. Some time in the eighteenth century it finally became clear that the different components in the air had to be distinguished. One of these components even involved fire—another element from the classical age. The issue at hand was combustion. In the burning process, something was separated from the air and absorbed by the burning material. Lavoisier discovered this phenomenon in many experiments after 1770 that he carried out mainly with scales, impressively demonstrating the value of precise measurement.[7] The component of the air that Lavoisier separated, today we call oxygen.

Lavoisier would not have been able to make this discovery if he had not first formulated a new theory (known today to be correct) of combustion.

In the past, chemists believed that when something burned, it released a material called the fire principle or phlogiston; at first glance this is exactly the impression that burning makes. In this case, the most simple sensory observation deceives, but after Lavoisier reweighed the burned material under controlled conditions and found it heavier rather than lighter, his evidence disproved the idea of the fleeing phlogiston. An old hypothesis could now be falsified. It was his new idea, that of the existence of a new material supplied to the original material, that allowed Lavoisier to see something that had eluded other scientists for a lifetime.

Using scales for measuring allowed Lavoisier the perception that something might not be quite right with the old paradigm, as Kuhn would say, who explained the subsequently new view of a phenomenon using the familiar portrayal of a visual metamorphosis (Fig. 5.1) when he wrote, "What were ducks in the scientists' world before the revolution are rabbits afterwards. The man who first saw the exterior of the box from above later sees its interior from below"[9] (Fig. 5.2).

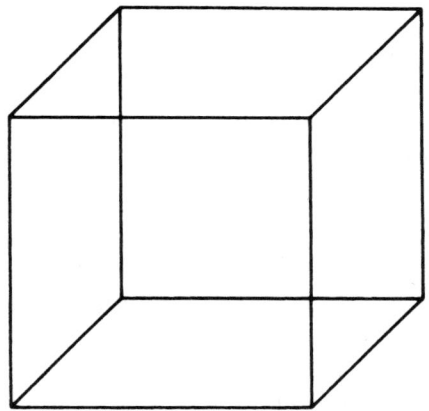

Figure 5.2.   The Necker cube, which can present a three-dimensional structure—a cube—in a two-dimensional picture. As observers, we can perceive the seen object, that is, recognize it as a spatial form in many different ways. If we see the cube at the "lower left" as being in front, the cube at the "upper right" moves back. We can also see them the other way around. What's important is that our perception of the figure is not up to us. If we look at the Necker cube longer, we notice that both versions keep going back and forth just like the ambiguous duck–rabbit figure in Fig. 5.1 (if we see and recognize both animal shapes in the first place). We can't see either image continuously; approximately every three seconds we have to switch.

Strokes of Genius and Revelations

Unfortunately, Kuhn just as soon afterward neglected his analysis of perceptual change in scientists to turn his attention to other questions, among them how scientists experience internal transformation. As we might expect, no one has thought to look for a logic of scientific discovery here. On the contrary! Because the subject is a "religious experience," the terms are "scales that fall from the eyes," as in the Bible story when Saul becomes Paul and a "flash of lightning" enters his consciousness and "through the grace of God" illuminates a previously dark riddle with the illumination again arriving in sleep.

As is so often the case, Kepler would have understood all of this and would have associated the flashing with inner images of the mind that converge with the perceived behavior of external objects. Essentially, the sudden inspirations, the new forms of perception in scientific transformation, suggest the influence of archetypes. To understand revolutions in thinking and grasp the process of science, we must trace the activity of archetypes; the idea of beauty is indispensable to this process.

Distinct Forms

The concept of form, which I have already discussed in terms of perception (Figs. 5.1 and 5.2), is vital. The pioneers of gestalt psychology in the early decades of the twentieth century already formulated the view that the mind is organized according to configured patterns.[9] The task of such an epistemology would be to find out more about the "gestalt perception" as a source of "scientific cognition" as Konrad Lorenz in 1959 did in his essay of the same name.[11]

The surprisingly enduring tendency of human perception to construct entities out of patterns and outlines in spite of all existing knowledge implies the lengths that perception goes to recognize beauty (Fig. 5.3). According to Hutcheson, the idea of beauty is triggered by a form and is then especially noticeable when the figure displays unity in variety.

The form that certainly achieves this unity in variety is the circle, and thus under aesthetic specifications it is not surprising that the shape of a circle is extremely easy to perceive. We can perceive it both in stationary (Fig. 5.3) and in moving pictures. If we roll a wheel in the dark that has

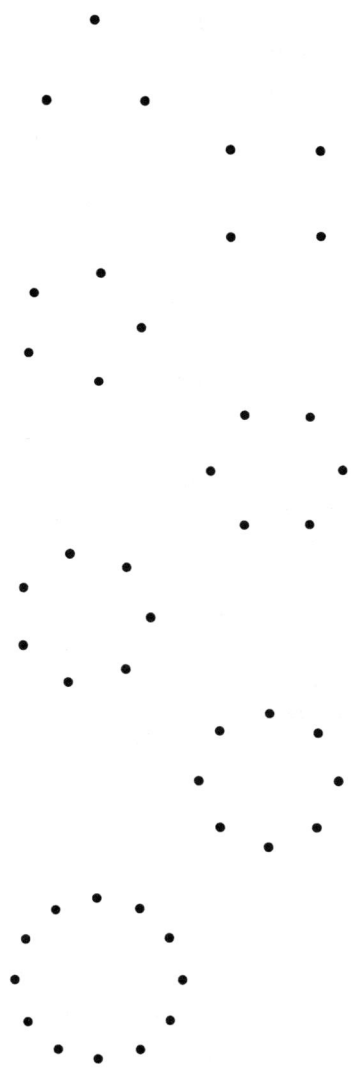

Figure 5.3.   The illustration shows two properties of gestalt perception. First, the corner point of the square appears to be connected with imaginary lines although there are no real lines. Second, for the observer the imaginary lines quickly turn into circles.

little shining bulbs attached on the circumference, we can see, with only one or two little bulbs, the line of motion that a single light travels on the wheel. Mathematicians call the resulting curved line a cycloid. This curve disappears from the field of vision when the number of bulbs exceeds six. The shining points then melt suddenly into the form of a circle. This is the form an observer wants to see, in darkness and in science.

Beauty in Paradigms

In my opinion, any attempt to describe transformation of scientific perception without the idea of beauty must fail. For us to analyze what changes when our perception is drawn to something that results in a new thinking style, we must take aesthetic conceptions seriously.

To be able to put this analysis in the framework of scientific discourse, we first need to define what is meant by a conceptual change in science, ever since Thomas Kuhn referred to this concept as a *paradigm* or *paradigm shift*. These words had already been used in philosophical discourse for some time, but their contemporary meaning is drawn from Kuhn's application, which has currently even found its way into everyday usage.

Kuhn's original view is presented in Table 5.1, which names some criteria to be examined in brief case studies from the history of science. As far as I know, Kuhn's model presents the general view accepted by scientific theorists. Kuhn's paradigm has also become a process itself, always a risky development. I cite Kuhn's concepts from 1977 in Table 5.1.

Accuracy

We expect a certain amount of accuracy from a scientific description (theory) of reality, and Ptolemy, Copernicus, and Kepler alike strove for accuracy when they developed and published their conceptions of the

Table 5.1.   Criteria for a Paradigm for the Operational Framework of Normal Science

Accuracy	Fruitfulness
Consistency	Simplicity
Breadth of scope	

cosmos. Although each of them wanted to make predictions as accurate as possible about particular celestial constellations (solar eclipses for example) or planet conjunctions, the unreliability of the geocentric stance certainly contributed to the discarding of the Ptolemaic system.

Wasn't there something more important than accuracy? More persistent, and thus more paradigmatic than the negligible precision of observations without a telescope, was the idea of the circular orbit. In astronomy for many centuries it was self-evident that planets moved in a circular orbit. In fact, the circular orbit was the paradigmatic core of the early teachings of astronomy, which were already referring to the concept as archetypal. It was only after 1600 that with increasingly accurate measurements (as before, without a telescope) it was becoming clear that Mars, for example, revolved around the sun in an ellipsis. Tycho Brahe had been collecting the corresponding data, and after Brahe's death in 1601, Kepler understood the full extent of these data. In the course of the seventeenth century, Kepler showed that he could more accurately explain not only one but all planetary orbits if he described them as ellipses (Fig. 5.4). He commemorated this event with his three laws. This was the only way he could finish what is known as the Copernican Revolution and subsequently launch a new image of the cosmos. Kepler made the paradigm shift demanded by the increasing accuracy of measurement. The beauty of the circular orbit that had to be sacrificed he recovered by viewing the motions of the planets as the image of the trinity.

As important as the accuracy of a world-view is, because its relevance frequently depends on technical provisions (hand telescope, space telescope), the idea of accuracy tends to pale in comparison with other ideas in astronomy—including those about circular orbits.

Consistency

We also demand of a scientific consideration of reality that it won't get tangled in absurd contradictions. When this happens, we have to abandon it altogether, or at least its assumptions in relation to Kuhn's second requirement of consistency. Aristotle had the seemingly logical and self-evident idea that heavy objects fall faster than light ones, but he failed to notice that intellectually he was getting hopelessly muddled. Only in the seventeenth century when Galileo incorporated Aristotle's suggestion into a new mechanical model (paradigm) was the inconsistency discovered. In

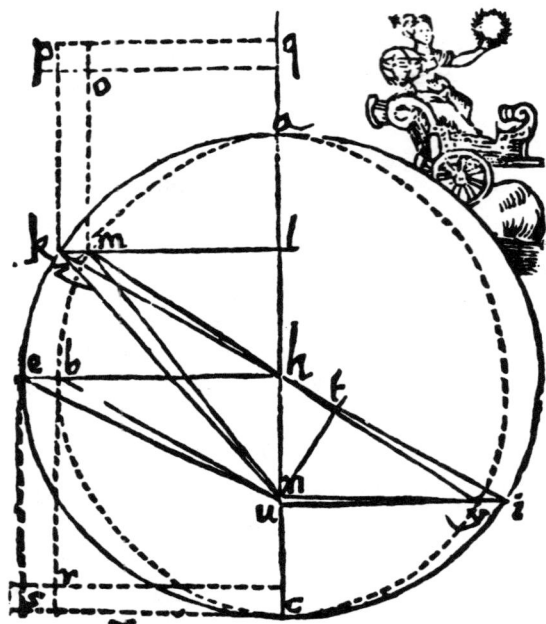

Figure 5.4.   The orbit of Mars is in fact an ellipsis, but we can't see this with the naked eye. The difference between the long and the short axis of the Mars orbit is only 0.5%.

a thought experiment, he imagined combining a light and a heavy body with a string (Fig. 5.5), and he asked himself and his students what would happen when the two bodies dropped. Would the slower, light body retard the heavy body or would the faster, heavier body accelerate the light one? The answer is of course that both objects fall equally fast, and on account of this inconsistency the old physics of Aristotle can no longer lay claim to validity.

Breadth of Scope

The third point of Table 5.1 shows that an explanation can only claim the scientific rank of paradigm when it is valid and helpful beyond the narrow context in which it originated. This idea of breadth of scope is demonstrated by the quantum mechanics of the atom, asserted in the first decade of the twentieth century and submitted in mathematically

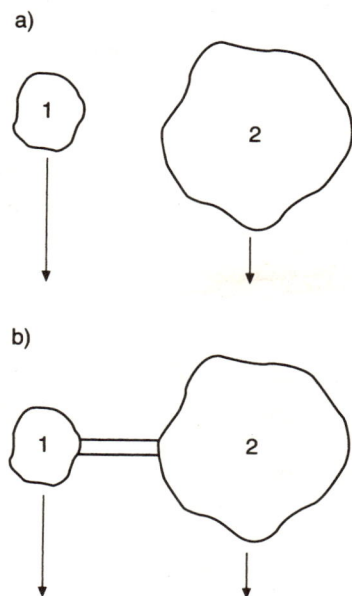

Figure 5.5. Galileo's demonstration of the contradiction in Aristotle's physics. When two bodies (1 and 2) free-fall, then according to Aristotle the body that is twice as heavy (2) moves twice as fast (a). Galileo used a thought experiment to show that this doesn't work. He proposed that both bodies be tied together (1 + 2) and asked how fast the combination of the two would fall (b). Obviously they can't fall three times as fast as the small body (1) because there is no new force. The only possible conclusion is that all bodies in a free fall move equally fast.

complete form in the mid-1920s. This new mechanics took over the classical version and was developed as a new theory of matter because without it, the stability of the components of matter—the atom—could not be explained. Quantum mechanics delivered after its completion, however, much more than the physicists had ordered. Without any kind of new assumptions, quantum mechanics put them in the position to show a mechanism for bringing about the chemical bond between atoms and molecules. With the quantum theory of hydrogen atoms, physicists obtained the theory of hydrogen molecules in the bargain. This meant that along with comprehension of the physics of the atom they also gained a comprehension of the molecule. Today it is believed that quantum theory furnishes even the most fundamental physical comprehension of all matter, including crystals and solids. Its emergence has created a new style of thinking, in fact the most exciting paradigm science has to offer.

Fruitfulness

Alongside the breadth of scope of a statement within a given physical formulation of a question—in the preceding example it was the stability of atoms and molecules—every paradigmatic theory attempts to yield results far beyond the borders of the discipline. Fruitfulness is introduced by Newton's theory of gravity. The force of gravity that the English physicist understood as a young man is valid not only on earth and in the sky, and it applies not only to the moon and to an apple, it is applicable on all planets and for all terrestrial objects and is thus fruitful, or fertile, in more than one respect.

I should do a little more justice to Newton's theory; his mechanics celebrate their great triumphs because they deviate from Kepler's laws in that they can be explained by a single operative force—gravity. Gravity was simpler than all of the available explanations of the world up to then and it is probably this quality of Newtonian theory that has left behind the greatest impression in the long term. The effectiveness of the ability of a paradigmatic theory is shown in that in his own observable form, Richard Feynman repeated, in a lecture, Newton's chain of thought.[12] After Feynman's death (in 1988), a great effort was undertaken to find the believed lost "Motions of Planets Around the Sun" lecture on why planets move elliptically—"Feynman's Lost Lecture"—in archives and to make it available to the contemporary public in book form. Part of this effort can be attributed to Feynman's appeal alone. It had foremost to do, however, with understanding the impressive influence of Newtonian mechanics and the associated world-view.

Simplicity, Part I

The key word *simplicity* is not just the last word mentioned in Table 5.1. With this concept for explaining the development of science in the paradigm model, Kuhn's entire structure begins to wobble a little. The criteria discussed thus far all have something to do with what we might call the correctness or truth of a theory or of a scientific statement. The fifth point simplicity deviates, however, because the connection this criterion has with the accuracy or correctness of a theory remains unclear and insufficient. What is meant in terms of Newton's mechanics, of course, seems obvious. We can certainly say that Newton's one law of gravity is simpler than Kepler's three planetary laws. If we want to be picky, we can add that Copernicus's explanation for motion of the planets got by with fewer auxiliary constructions (epicycles) than did Ptolemy's explanation. Both Kepler's and Newton's laws are apt, but as far as Copernicus and Ptolemy are concerned, both are actually wrong, technically speaking, because neither the earth nor the sun is in the center of the universe. Additionally, Copernicus and Ptolemy are only marginally distinguishable when it comes to the number of epicycles each had to add to the circuits of the celestial bodies to bring the observed circular motions into line. That they both erred in the same point only serves to emphasize their attachment to simultaneously simple and beautiful circular images.

The very powerful criterion of simplicity is apparently related to a different quality than the four previously named points in Table 5.1, and the suggestion has been made to create two separate lists of criteria for scientific theories or their styles of thinking (paradigms) to understand what people find interesting about them. Indeed, theories might, on one hand, be obviously correct (and possibly even true) and afterward rise to the level of a paradigm that would rule over the scientific thought of a given time. On the other hand, the influence of available explanations also depends on whether they are perceived as beautiful and aesthetically pleasing, and simplicity is not the only criterion that might lend them this quality. As I tried to demonstrate in the first chapter, theories are very highly valued when they stick to images that are instantly appealing.

It's therefore worth distinguishing between two groups that can define a theory or a paradigm. They are compiled in Table 5.2, and this overview of a suggestion can be traced back to the scientific philosopher James McAllister teaching in Holland, who devoted an entire book to the connection in *Beauty and the Revolution in Science*, published in 1996.[12]

Table 5.2.   McAllister's Criteria

Criteria for "truth"	Criteria for "beauty"
Internal and external consistency	Simplicity
Predictive scope and accuracy	Symmetry
Fruitfulness	Analogical interpretability
	Metaphysical consistency

Table 5.2 is a first attempt to represent the progress of science more accurately than with the simple system à la Kuhn. It should in no way be understood as a complete consideration of all categories involved when the scientific paradigm changes, and along with it, the understanding of the world.

If we use this list to define what makes up the aesthetic charm of a scientific notion—such as the theory of relativity—then the answer might be that we prefer one scientific description of the world to another equally correct one when it is as simple as possible, it contains symmetrical elements, we can observe the explanation through a model that is accessible and gives an overview, and we do not have to choose between other world-views, for instance, the Christian faith, to accept the theory. The following is an attempt at making these points discernable.

Simplicity, Part II

I would like to start out with the aesthetic category of simplicity, reminiscent of Hutcheson's suggestion that beauty is concerned with the uniformity that finds a theory in the variety of reality and makes it visible. Simplicity is indeed anything but simple (in the sense of a simpleton) and is therefore probably so much more important for the acceptance of a paradigm or style of thinking. Simplicity is found, for example, in the Newtonian laws that alone can make the claim of being beautiful.

Before I say more about this form of simplicity, a concept that fills entire volumes without being exhausted, I would like once more to remember the other kind that has already been introduced in connection with Nicolaus Copernicus. When he suggested, in the sixteenth century, removing the earth from the center of the universe and putting the sun in its place (heliocentric system), he also tried to find out whether he could explain the motion of the planets more simply than Ptolemy. "More simply" is meant here strictly quantitatively, as already noted. Copernicus wanted to manage with fewer epicycles to explain planetary motion that everyone had thought of as circular. This applied only conditionally and as a result the sun remained in the center because it was bound together with entirely different aesthetic and theological notions.

In the twentieth century, it was Einstein who again and again valued the higher form of simplicity in his fundamental assumptions and to some degree understood it as one of the cornerstones of his theories. Einstein's special quality consisted of seeing "simple connections" in events that seemed to be going in completely different directions. This special quality also consisted in thinking through his decision logically to the end, until it was firm, regardless of how much energy it took year after year. His early decisions in favor of aesthetic criteria reminiscent of Jacques Hadamard gave Einstein the strength to keep from getting lost among all the rational solutions and mathematical transformations, to finally get what he desired in the first place and what was to be expected—a beautiful theory.

Simplicity, Part III

I will make one last comment regarding the difficult idea of simplicity. The historical conflict between Goethe and Newton that revolved around the nature of light and color has its place in this context by demonstrating how difficult the application of an aesthetic criteria can be. Simplicity meant something completely different for Newton than it did for Goethe. For Newton it was light perceived as color, which he observed after the refraction of a light ray through a prism, simply because it couldn't be broken down any further. Blue light, for instance, can be characterized by a wavelength and differs in composition from sunlight. Blue light is therefore simple—for Newton. For Goethe, the blue light coming through Newton's device is complicated because he can only see it by using a special gadget (a prism). Sunlight, on the other hand, is simple because it simply exists in nature.

It is easy to see that both standpoints, complementary to each other, can be justified. It is just as easy to accept that they belong together even if at first glance they rule each other out, if we mean to fully understand the nature of color. It is, however, apparent that the preference we can develop for the description of something such as light does not depend on the correctness or the degree of truth of the description, but depends more on the pleasure the premise offers. We make decisions based on our tastes and choose the one we find most appealing.

Symmetry

The concept of symmetry is discussed in detail in the next chapter, so I only refer to it briefly here. Symmetry—or harmony—or regularity—recalls the ancient commitment to beauty we can trace back to the Pythagoreans, who believed that beauty is determined by the correct agreement of parts among themselves and as a whole. Symmetry was Einstein's main

requisite for the equations and laws when he was disturbed in 1905 by an
asymmetry lurking in classical electrodynamics and its famous Maxwell
equation (having nothing to do with the actual appearances). Faraday, on
the other hand, revolutionized physics in the early nineteenth century
when he worked long and hard to find the symmetry between an electric
current and a magnetic field until he could prove it (and use it) beyond a
shadow of a doubt. It was well known that an electric current could deflect
or influence a magnetic needle. Faraday's study implied that a magnetic
field should then also be in the position to change an electric current or set
it in motion (and in fact he could confirm and implement these thoughts in
1831 when he discovered electromagnetic induction as it is called today).

Observable Models

Important for the acceptance and influence of any theory is its analog-
ical interpretability or potential for representation by a graphic image.
Similarly, the observable model applies to René Descartes's seventeenth-
century image of the universe filled with whirlpools and how, also on
account of its accessibility, his notion held scientific minds prisoner for a
long time. Just having to work on an observable concept allowed Newton
to postulate the so-called long-range effect that led to his mathematical
version of the law of gravity. Finally, the "psychologically most difficult
piece of work" Einstein had to pull off consisted in doing away with ether,
considered to stretch across the expanse of space, and to replace it with a
type of free field that would ensure the end of empty space.

Metaphysical

We can also mention Einstein in connection with the harmony of
world-views: Einstein addressed the metaphysical components of a theory
when he refused to acknowledge the random nature of atomic reality in
the famous reproach, "God doesn't play dice." That a lover of intellectual
freedom in all other respects could make this comment reveals the need for
a determinism for natural laws that constricts us but in return offers a form
of intellectual security. Einstein conducted his science very often like a
small boy who is looking for things hidden away by his mother, Mother
Nature in Einstein's case. He relied on the knowledge that by using his
talents he could find them.

Einstein may have had difficulty with the physics of the atom deriving
from a quantum nature of reality because the unstable quantum stood in

direct contrast to the continuous field in which Einstein saw the physical world embedded. This introduced a fragmentation into the whole quantum theory which expressed that no causality existed in relationship to the individually observed specific case. The premise that, even beyond this, the whole theory of matter rested on statistical foundations, Einstein would not and could not fully accept.

With his refusal to confirm quantum theory, Einstein touched a sore spot of physics and made crystal clear that science could only account for the general and overlooked the exceptional specific case. However, this fragmentation didn't mend the way he had wanted it to, with a return to classical methods and an all-embracing unified field theory. If we characterize as rational all that a scientific theory can explain, then Einstein's criticism of quantum theory called attention to the fact that there are irrational elements of reality, meaning the individual events that cannot be deduced. If we look at it this way, it is the irrational that is unreconcilable to a great mind. In fact, it is difficult for most people to find irrational elements appealing or satisfying.

Aesthetic Transformation

As indisputable as it appears to be that aesthetic moments are integral to profound scientific transformations and that scientists quite simply prefer certain images and models to others based on their beauty, it is indisputable that scientists' opinions of what constitutes beauty can change. Thus, the first astronomers to hear of Kepler's ellipses were not altogether pleased that there were now centripetal orbits. Ellipses were considered warped and imperfect circles.

This attitude changed, however, with the empirical success of the new notion of orbits—the new astronomy. A clear example is the mathematician Peter Crüger from Denmark who, as a contemporary of Kepler, in 1624 expressed serious misgivings about an elliptical orbit around Mars, but who five years later not only came to terms with it, but even reached the point of finding it beautiful. Kepler's theory and the heliocentric system in the heavens were so well constructed that it was suddenly possible to see the beauty of an ellipse.

Similarly, the success of quantum theory caused what is called an "aesthetic induction." The unobservable nature latent in the new atomic mechanics and their subsequent abandoned search for universal causality at first not only bothered many scientists but "sickened" them, in the words of Viennese physicist Erwin Schrödinger when he heard Heisenberg's

Figure 5.6. The observable image of an atom that Erwin Schrödinger had in mind when he developed his version of quantum theory.

description of the atom for the first time. Schrödinger put everything he had into developing a countertheory that would retain at all costs what he required of a physics theory—observability. When even the electron that surrounded the nucleus in a hydrogen atom couldn't be said to exist as a particle of an orbit, then it should at least be able to be calculated as an evenly swinging wave (Fig. 5.6). Critics consider Schrödinger's thoughts, mathematically conveyed, an elegant theory of matter, and they prize his equation as a work of art in perfect (mathematical) form.

Schrödinger's starting point for his wave theory of the atom was a hypothesis set up by French physicist Louis de Broglie in 1924. It seemed strange to him that only light had a dual character and could appear not only as waves but also as particles, as Einstein explained in 1905. De Broglie asked himself why no one saw the asymmetry that now existed between light and matter. What made sense for one physically fundamental form of reality had to be true for the other as well. He therefore hypothesized that there must be not only light waves but also matter waves (and he would soon be proved right—verified).

Of course de Broglie knew that physicists could specify a mass of electrons and had no idea how to introduce an electron as a wave, but the aesthetic argument of symmetry won the upper hand for him, so much so that he dared expose his idea to the public and maintained that even electrons and other particles had a wavelength. He did this "solely for the sake of intellectual beauty" and was not particularly surprised when nature answered by confirming the symmetry.

6   *The Distinct Form of Symmetry: A Look in the Mirror and Other Operations*

Mirror, mirror, on the wall,
who's the fairest of them all?
Snow White's stepmother

The time has come not only to understand how beauty can make science possible but also how science can explain beauty—in the not too broad context served by a biological or a physical method of presentation.

In the next few chapters, we are also concerned, scientifically speaking, with the form of beauty that is more generally familiar than the beauty of science. I am referring to the beauty we experience when we glance at a face or a body, and I am also referring to whatever biologically sound reasons exist for the discrepancies in the perceived beauty of certain forms—why we see some forms as more attractive and others as less attractive. Symmetry provides a good place to start the inquiry because even the simplest possible pattern gets the benevolent attention of an observer just by being symmetrical.[1] We can see this by splashing paint on a piece of paper and folding the paper so a symmetrical spot forms; already we find the image appealing.

Symmetry is most evident, as in the case of the paint spot, in the form of the mirror image. This concept even plays a role in physics, with the observation of the smallest objects of this science, as well as in psychology, with the observation of other people.

Face and Body

The faces of people and their bodies both attract attention based on their symmetrical quality, or at least on the degree to which their proportions

113

aspire to being well balanced. However, there seems to be an essential difference between the two and thus a type of total asymmetry, because "we are in the habit of wearing clothes" as the eighteenth-century Giacomo Casanova regretfully maintained on a regular basis and went on to wonder that "the face we willingly expose for everyone to see contributes the least to our own personal gratification."

It was Casanova who made the observation that the face of a woman is of little concern when the lights are out (or maybe not). He went even further and remarked that the face of the one seduced is the deciding factor in whether the seduction turns out to be gratifying. This was the famous Venetian's favorite topic, and he led us into the contemporary scientific realm of evolutionary biology by asking, "Why must it (the face) play the starring role? Why do we fall in love with the face? Why do we judge the beauty of a woman only according to this single testimony, and why do we excuse her if her covered body parts don't complement her pretty face?"

I am not entirely convinced that Casanova is right. I am, however, familiar with the reverse situation because again and again it is apparent that even the smallest deformity in a face—even the slightest cross-eyes, for example—can deny a person the attribute of beautiful even if he or she has an impeccably or gloriously built body. Thomas Mann depicted this relationship in the novel *Joseph und seine Brüder* with Lea, the older sister of Rahel.

Many men and women today do judge the beauty of a person by the face alone, as Casanova suggested. Before I can try to put this difficult question in the context of evolution, I have to explore if we even know what leads us to consider this person's face beautiful and that facial expression ugly. Can we establish unequivocally when a face will be perceived as beautiful by as many people as possible? Do biological aspects prevail or is it a matter of cultural differences?

Anyone undertaking this subject can learn a great deal from both asymmetry and symmetry; it is worthwhile therefore to become familiar with both aspects.

Symmetry in Science

The concept of symmetry derives from the Greek word *symmetros* formed from a matching syllable—*syn* became *sym*—and the root word *metrein* (measure) and combined them to mean something like "to measure together," which can be better translated as "proportional." This concept refers to the mutual correspondence of parts, such as the relationship of size to form. Because in most everyday activities—those outside scientific jurisdiction—symmetry is used to refer to parts of a human body or to the halves of the accompanying face, and because it is apparent that the right and the left sides of the body, seen outwardly, are more or less equally built, we often use the concept of symmetry in the sense of the mirror image. That things work differently internally and that the two halves of the human brain are expressly asymmetrical I discuss further on.

When I talk about visible symmetry in this chapter, I am referring, if I don't mention otherwise, to the equality or balance we either recognize or miss in the mirror. In science we have greatly expanded the concept of symmetry even to the effect that we say an object (or a state or a law) possesses the quality of symmetry, when we can indicate a well-defined operation (a transformation or a variation) performed on the object (or the state or the law), so in the end we get the original form, and the new situation cannot be distinguished from the initial one.

The concept of operation here is only related to the surgical term as far as it is understood concretely, and it expresses that an object is changed (or its position is) so that afterwards everything is as it was before, namely, everything is in order or in good health. Mirroring is an operation; when I see my face in the mirror, I am observing it after the operation of mirroring. A rotation can likewise be thought of as an operation: When I rotate a square around a right angle (at 90°), it looks, after this symmetry transformation, exactly as it did before. In a similar way a six-sided (hexagonal) snow crystal (Fig. 6.1) doesn't change when I turn it around the sixth part of a circle (60°).

The six-sided snow crystal and its symmetries preoccupied Kepler, who by 1610 realized that none of the many snowflakes he examined quickly before they melted in his hand had five or even seven spokes.[2] Each time, he found six points of the "little snow stars," as he tenderly called them, and a scientist asks in such a case what the cause of this regularity might be. Without coming to an answer that from the modern standpoint would be complete (modern scientists talk about molecular structures as yet unexpected and inaccessible and they mention lattice forces when they talk about the physics of crystal formation), Kepler acted according to the conviction that "the six-cornered form" develops out of necessary adherence to laws concerning matter or from its own nature because "the prototype of beauty" dwells in the six corners.

Figure 6.1.   A few examples of snow crystals.

The prototype is the previously mentioned archetype. Kepler expressed that the pleasure human beings take in regular forms of natural things is not a coincidence. Their conformity and their beauty derive from one source: the archaic level on which the archetypes have settled. They form both mind and nature and their harmony is consequently judged according to aesthetic criteria.

A First Law of Conservation

Symmetries not only have statistical but also dynamic consequences. Indeed operations can help present many regular molecular or crystal structures and their symmetrical properties; in spite of the visual charm, however, I do not intend to investigate further. Symmetries in physical processes and fundamental laws are far more interesting. Mathematics has helped show that the fundamental quantities that survive in the world of physics, the uniformity that survives among all its varieties, are associated with symmetries.

When an aspect remains unchanged under a certain operation, it is awarded the property of invariance. For example, we say that a snow crystal is invariant under a rotation of 60° and so we notice that regular six-cornered images of this kind are also symmetrical mirror images. I can turn it 3 times 60°—180°—and exchange right and left.

Now there are a few accessible forms of symmetry or invariance that at first glance seem trivial or even banal and seem to deserve no special attention. These forms, however, prove fundamental for physics. I am referring to the statement, apparently needing no proof, that the result of a scientific experiment should not be dependent on the time the scientist starts working and making measurements. In other words, the laws of physics should be invariant in terms of fluctuations (called transformations) in time, when time is understood as what a clock shows. This designation is sufficient for physics.

The beauty is now that, in the context of classical mechanics, this transformation invariance produces the possibility of proving through mathematics that the energy of a system survives independent of its appearance or its number of permutations. Concretely, this means that there is no change in the total energy (over the course of time) called for in the first law of thermodynamics. The conservation of the puzzling quantity energy is the consequence of symmetry in reference to time.

I'd like to make a personal remark here: The most extremely elegant inference of this conservation law (and of the other two laws connected with it) is found in chapter one of the first volume of a famous *Course of Theoretical Physics* that comprises ten volumes. I am referring to the work by the two Russian authors Lew D. Landau and Iljy M. Lifshitz. Generations of physics students have learned their theories from Landau and Lifshitz and hopefully, after they have mastered the tool, have allowed themselves to be gripped by the admiration that is traditionally shown only for buildings or works of art. Landau and Lifshitz have very likely inspired many students to pursue physics on account of its inherent beauty, and this is wonderful when we consider that theoretical physics as a science has achieved the highest level of abstraction. After distancing itself eons away from the general awareness, it finds its meaning once again in the long-abandoned perceivable beauty of nature. What we understand with the senses directly seems to appear as well at the heights of theoretical speculation. Beauty is everywhere.

In the most recent edition of the first volume there is a supplement on the life of Lew Landau, who I presume to be the intellectual father of the textbook. There, the reader learns how Landau is able to present physics so ingeniously. He talks about how he was personally deeply influenced by the unbelievable beauty of the general theory of relativity and speaks in connection with the quantum theory of the pleasure of true scientific beauty and of the great rapture that came over him when he reflected on it with formulas.[3]

It is important to continually make clear that the great physicists of our time, Einstein, Bohr, Heisenberg, and others, found this beauty in their science and immediately fell under its spell. One of many sources of admiration is the connection between symmetry and the laws of physics and the conservation of fundamental aspects (invariance). Thus, beauty manifests itself precisely in the way the Scottish philosopher Hutcheson described it, as uniformity among variety. Uniformity even appears twofold. It appears once in the law of conservation and again in the fact that it always rests on a founding principle—the foundation of symmetry.

It was a woman who discovered this far-reaching connection, Emmy Noether, a mathematician from Germany who eventually emigrated to the United States (1882–1935). For the first time, in 1918, she combined abstract symmetries (invariances) and concrete conservation laws of physics and thus gave science the chance to be beautiful.

More Laws of Conservation

Besides the law of the conservation of energy, Emmy Noether was concerned with the laws of momentum and of angular momentum. To begin with momentum we must think not of displacement in time but of a linear displacement in space. We can repeat *mutatis mutandis* in this instance because it is just as true that the outcome of an experiment is not dependent on the translation of the equipment parallel or back and forth (without turning it in the process). In other words, the laws of physics must be invariant not only in relation to a time but also in relation to a spatial translation, and this symmetry results in the conservation of momentum, if we remember that momentum is defined as a product of the mass and the velocity of a physical body.

We can take yet a third step and add that the arbitrary rotation of a system in space changes nothing about the mechanical properties of a system (after the operation is over). We can also derive the conservation of a quantity; this time it is momentum that a system possesses that does not move linearly but rotates. We speak of its angular momentum and know now that even this survives.

The Scientific Look in the Mirror

We should not get the impression that everything is symmetrical and all symmetries that we can think of or easily imagine are relevant to our discussion. It is extremely important to make clear, however, that physicists know as much as they do about atoms and smaller (fundamental) components of matter because these elements of reality reflect basic space–time symmetries in an elementary way—in mathematical form. Indeed, the concrete, observable quality existing on the level of material (macroscopic) reality is lost. However, the inner observable quality survives in the idea of symmetry. Atoms can thus be grasped aesthetically, and why shouldn't they be?

With the inclusion of the microworld in the network of symmetry there was the following surprise: Physicists—with quantum theory already behind them—searched for decades, well into the mid-1950s, for the natural laws at the heart of the universe on the assumption that it had an exact symmetry that occurred threefold. In each case of symmetry there is a kind of reversal or reflection: first with a right–left exchange (with a reflection of space), second with a change of the electric charge (meaning with a change of its plus sign (+) to minus (−) or vice versa), and third with a reversal of the time direction (which then runs backward instead of forward, which in these dimensions is actually imaginable as is explained shortly).

It is easy to imagine the universe remaining unchanged if we exchange positive and negative signs of electric charges. First, the existing designation—we'll call the electron negative and the proton positive—took place only out of convention, meaning we could just as easily call them the reverse. Second, a negative charge has the same effect on a positive charge as the other way around. The same force appears in the same direction. No one worries about this symmetry, and no one questions it even halfway seriously. We use it to make the laws of physics comply with the conservation of charges.

With time reversal, on the other hand, it gets trickier to imagine—no one actually takes it seriously in relation to ordinary events. In our everyday world, clocks run forward, and time travel into the past is out of the question (otherwise someone could murder the father of my wife before she was conceived). However, this symmetry is not meant in such a simple way. Our lives are in no way time-reversal invariant, but the movement of an atomic component is—at least the laws of nature allow for it. In essence, the fundamental equations of physics know no time direction, a fact conceded without debate and one that presents scientists with the enormous problem of finding an explanation for the observable direction of physical time. Its arrow flies only ahead and we suppose that it (the arrow) is only produced when many events concur and we can use the concept of probability. Only something happening in the future can be called probable. In the past it has already occurred (or not) and statistical assertions make no sense. Yet I am not referring to this macroscopic symmetry when I discuss the elementary realm of matter, where at the least it is possible to allow theoretical atoms and smaller units to run backward in time.

The main focus of our attention should be the symmetry I mentioned first: the space reflection also known among physicists as parity, or "equal." In the 1950s, to the general surprise of all, scientists discovered a fundamental asymmetry in nature. As strange as this may sound, the assumption that the image of a physical process we see in reflection is also a physical process possible according to the law of nature is not entirely correct and consequently false. We understand this, for example, about radioactive decay, which for historical reason we signify with the Greek letter β (beta). The force responsible for the β-decay of an atom, of cobalt for instance, does not comply with the invariance of parity (Fig. 6.2). In other words, the mirror image of the depicted decay does not exist in this universe. When we see β-decay, so to speak, in the mirror, we have before us something that does not exist.

The crazy part, however, is that in spite of nature's asymmetry under these circumstances, nature behaves 99.99% of the time with parity, meaning that the great majority of nature remains beautifully symmetrical and right and left are not differentiated. Only an expressly tiny portion of its influence favors the left side, providing the joke that God is "a weak lefty" (without yet knowing what He is hiding in the hand behind His back). We use the attribute "weak" because the force responsible for β-decay is defined as a weak nuclear force, which would lead us to understand that a stronger form exists—the strong nuclear force responsible for the cohesion of the atomic nucleus.

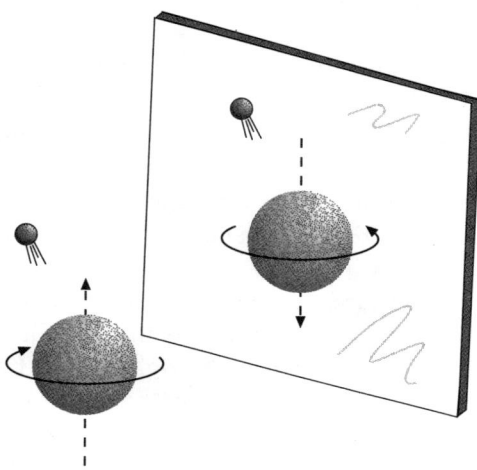

Figure 6.2.   The β-decay of cobalt and its mirror image. To be more exact, the cobalt isotope characterized by the ordinal number 60 is shown decaying. The cobalt nucleus is depicted as a ball. Physicists are aware that this nucleus has a spin that is (not easily) viewable as inherent angular momentum. It's important that the direction of the rotation allow a spatial orientation of the nucleus. It's also important that the mirror show an exact reversal of this direction. In the experiment, low temperatures were used to align the cobalt nuclei so the overwhelming number of them, like the depicted nucleus, point upward. Next comes β decay, meaning that electrons leave the nucleus. When they leave, as can be measured, they favor the direction indicated, that is, along the axis defined by the rotation of the nucleus. In the mirror image they fly upward, and this doesn't fit in with the reverse sense of rotation. This means that the mirror image of the β-decay of cobalt is out of the question physically. In reality it doesn't occur.

Why this tiny infringement of mirror image symmetry in nature exists and whether this subsequently implies a comprehensive asymmetry of the universe remains a book with seven seals for the physicist. There's nothing beautiful about it to them; in fact their astonishment at this deviation from perfect harmony endures. It endures because scientists do not rest content with a technically correct understanding of a particular state. They would also like to come to an aesthetic understanding of it. Dissonance still occurs between the known result and the expected experience. In my opinion, we keep searching either until we find another beauty or until we recognize what we find as beautiful.

The Everyday Look in the Mirror

The look we take in the mirror can do more than show us things and processes that do not exist without it (because they are ruled out by physical laws). When we look in the mirror we also sometimes see something we would not recognize without it: ourselves. It is difficult for us at first to learn that it is our own selves we see when we look at our reflection, and it continues being difficult for us to get used to the face staring back at us.

We know that our faces are not completely symmetrical because everyone looks different in a photograph than in the mirror. We look different because in a mirror image, the right and left sides appear reversed. I have become so used to my face in the mirror that when I see my image in photographs, it generally strikes me as a poor resemblance. This then leads me to the question of whether it might be true that a face that is perceived as beautiful in the mirror might not leave behind the same impression in reality (or in a photograph).

The exchange of the right and left halves of the face has consequences not only in terms of perspective, but also in terms of aesthetics. If we make a face (or the corresponding picture of one) perfectly symmetrical by creating it out of two right or two left halves (Fig. 6.3), we see that observers consider the "right" face more beautiful.[4]

In completely asymmetrical faces, in fact, what we see going on in the right half of a face makes the strongest impression. This is demonstrated by a picture in which the smiling and the serious-looking man's face are divided and put together anew in such a way (in genetics this is called recombination) so one time the right half smiles and another time the left

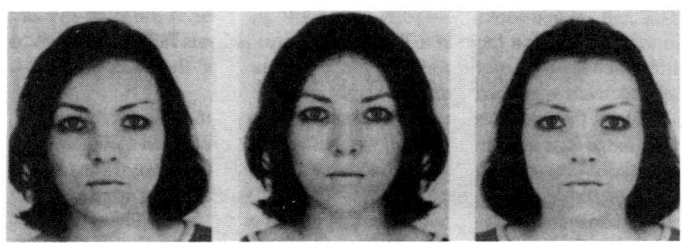

Figure 6.3.   A normal face (on the left) and its recombination in the form of two right halves (in the center) and two left halves (on the right). In empirical studies 90% (!) of all those asked considered the "right" face the most beautiful.

Figure 6.4. Which face looks happier and which looks sadder? Right-handed people think that a face smiling on the left seems more cheerful.

half smiles. We can also demonstrate this finding with a simple drawing (Fig. 6.4). When we have a happy face, the look we take in the mirror mutes this impression, and we see ourselves as more serious than we actually are. When the open and inviting smile—for example, on a woman's face— looks beautiful, then a potential suitor might want to avoid looking at her in the mirror if he intends to fall in love with her looks.

The Symmetrically Average Face

What is surprising about this experiment is the observation that a symmetrical face leaves behind the impression of being the most beautiful. This observation is valid in general and was made the first time—however unwittingly—around the end of the nineteenth century.

In 1878, the English scientist and anthropologist Sir Francis Galton— who is always referred to as the cousin of Charles Darwin and one of the forerunning philosophers of eugenics—tried to devise an experimental law that would prove scientifically his notion that every attribute can be inherited (the positive as well as the negative). Galton believed, among other things, that individuals could be born criminals, and that their illegal leanings would be visible on the tips of their noses. For this purpose he used the brand-new technology of photography to ascertain the typical face of such a person and superimposed images of lawbreakers not in the sense of police detective work, but quite literally. Galton then averaged the faces. He produced what he believed to be the average face of a criminal (Fig. 6.5), and he hoped to be able to see the evil intentions directly.[5]

Yet to his annoyance Galton had to admit that he liked what came out of his manipulations—he did not find the average criminal frightening,

Figure 6.5.   Francis Galton's "Average Face" (1878).

but (on the contrary) handsome. To his further surprise, the expert on inheritance had to admit that the mixed faces became increasingly beautiful the more pictures he melted into one image.

I regret that neither Galton nor other biologists and natural science researchers of the nineteenth century explored his conclusion, published in the well-known journal *Nature*, and so it was more than 100 years later before science began once again to discuss the beauty of the average created by this method. In 1979, the American anthropologist Donald Symons finally seized on Galton's idea and made the curious suggestion that what people call beauty could be defined biologically by how average the face is.[6] Symons found the corresponding reason in biological evolution. Ultimately, he believed, we could not expect the development of life to tend toward the extreme. Natural selection favored the life forms with average qualities much more because these were best at finding their niche. The offspring of females with average qualities, in the animal kingdom, have the greatest chances of survival. If we apply this to people, then the men who propagate most successfully are those who find the average, run-of-the-mill women beautiful enough to become interested in coupling with or in marriage with them, progressing to the seductive act and conceiving children. For this reason evolution is also to have furnished people with the perception of beautiful when they see a face of average proportions.

Symons's hypothesis that selection pressure leads to seeing the average as beautiful was tested in the early 1990s according to Galton's methods, with the only difference that in modern institutions the composed faces were generated with a computer. The investigation revealed that most of those asked once again found the averaged face more beautiful than the individual pictures that had gone into making up the experimental exposures. Thus, the attractive (aesthetic) value of the average was confirmed scientifically.

Many psychologists (and other scientists) continue to feel uncomfortable with this conclusion and believe that somehow we have overlooked something. The average does not really look all that beautiful, and if we look at the faces belonging to real people who were considered beautiful in their time—Greta Garbo for example, or James Dean—then it soon becomes clear that their beauty cannot be attributed to the symmetry that makes the run-of-the-mill face stand out.

In fact, psychologists understand that there are at least two components that contribute to the beauty of a face, and to this end they have made use of one of its simple tricks.[7] First of all, just like Galton and his followers 100 years later, they used a computer to help build an average face (Fig. 6.6a) that was perceived as beautiful (and is).

This first average face belonging to "Ms. Mix" was composed from 60 individual faces also rated by those questioned. Subsequently, yet another average face was constructed out of the 15 faces that had been chosen as most beautiful. The result was "Ms. Beauty." The average of just a few beauties is more beautiful than the average of all the pretty faces (Fig. 6.6b). More people find Ms. Beauty pleasing than Ms. Mix. Why is this true? What is the deciding factor? The cheek bones? The distance between the eyes? The distance between the nose and the chin? The form and placement of the nose?

We can reasonably say that we'll never know. For ultimately, nothing particular about the face is beautiful; it is the face as a whole that is beautiful. A computer makes many manipulations possible, so the face of Ms. Beauty can undergo even further electronic adjustments. Ms. Beauty can be made even more attractive when the differences between her and Ms. Mix, as the distance between the upper edge of the lip and the tip of the nose, are emphasized just a little bit more and produce a third face, a "Ms. New." In the end, the scientific beauty contest is decided in favor of Ms. New (Fig. 6.6c).

The objection that this choice has little to do with science because the research subjects are drawn only toward the beautiful faces who smile at them from the front covers of magazines or movie screens can be refuted in that even 3- to 6-month-old children made the same choices. When babies

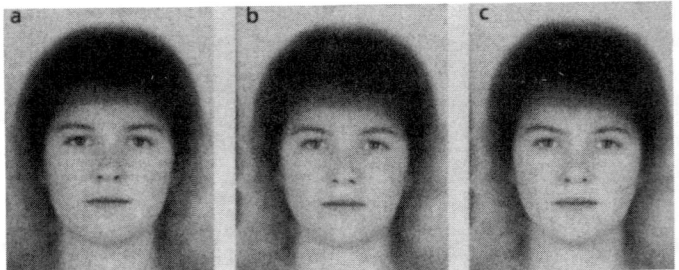

Figure 6.6. "Ms. Mix" (a), "Ms. Beauty" (b), and "Ms. New"; the text describes how they were made.

are shown pictures with faces that are considered beautiful by adults, the children pay attention to these pictures perceptibly longer than they do for photographs of people classified as less beautiful. The difference between a more and a less attractive face is by no means only acquired through cultural factors. It is determined far more by biology and so is something everyone can understand.

A further observation points in the same direction. Indeed, up to now we have observed only European faces—biologists refer for historical reasons to Caucasians—but it was easy enough to conduct the same experiment with Asian faces—specifically pictures of Japanese people— and the results remain the same (Fig. 6.7).

Obviously there is a culturally dependent understanding as to what makes people attractive—beautiful. Unfortunately, in the end the puzzle of what makes up beauty doesn't bring us any closer to the answer. It does, however, make the scientific perspective more accessible and thus more

Figure 6.7. "Ms. Mix" (a), "Ms. Beauty" (b), and "Ms. New" (c) in the Japanese version; the text describes how they were made.

interesting. Francis Bacon was right when he speculated, in the seventeenth century, "There is no excellent beauty that hath not some strangeness in the proportion." Whether and how this "strangeness in the proportion" is open to a scientific explanation is debated in the next section.

Symmetry and Sex

The scientists who research millimeters and angles in the face and want to know them in exact quantitative terms would have to try in vain to discover the deciding proportions of beauty. No one believes that there is a computer in the human brain that calculates the data that the sense organs input and that then outputs the verdict, "attractive." If we consider seriously that it is not the parts (and their quantitative relationship to each other) that make up the beauty of the face, but only the face as a whole, we can't make a judgment based on observation (or targeted parts), but only based on perception of the whole. However, we perceive without calculation. Perception is found at the beginning of every realization but always goes beyond the details and aims at the whole that we then call beautiful, directly, and without calculation (in the computer that our brain commands).

The question of perception becomes even clearer in the next chapter when I explore the more comprehensive and exact evolutionary conditions that can lead to the emergence of beauty. The process of perception is simplified, for example, when the object focused on is symmetrical. The influence of symmetrical patterns on perception can be examined in a graphic design found in various textbooks on "perception" (Fig. 6.8). The graphic design is arranged to influence what we perceive as figure and what we perceive as background. Each visual perception brings a figure into focus against a background.

When asked to judge which pattern is the figure and which is the background, most observers choose not by color or by brightness; although black tends to move toward the front more than white.[8] The discovery depends more on the symmetry of the pattern. The human apparatus for perception is constructed such that the symmetrical easily excites our attention, appears more attractive, and in the end more beautiful.

What good does this do the perceiver on the lookout for symmetry? We already know that symmetry is involved in evolution. Behavioral biologists have observed in many species of animal that the female prefers

Figure 6.8.   If you let your eyes go from left to right in the pattern, what is the figure and what is background changes. The symmetrical comes forward; the asymmetrical goes back.

to mate with males who offer or display symmetrical patterns, where their left and right halves of the body correspond.

The same has now been observed with people. In one U.S. study, seven measurements (sizes) were quantified in tests according to right–left deviations—for example, the width of the feet, ankles (joints), and elbows and the length of the ears.[9] Asymmetry was then associated with the sexual behavior of the student participants ascertained through questionnaires. The results were surprisingly unambiguous.

"Symmetrical" men not only had their first sexual experiences a few years earlier than their "asymmetrical" competitors, but through the years they also found new sexual partners faster and with more frequency. The influence of symmetry on the women's figures had to be analyzed much more closely to understand its attraction (measuring women's bodies is always more complicated than measuring men's). However, symmetry is also somewhat useful even to them because, according to the questionnaires, they experienced orgasm more easily and more often when they spent a night or more with a man built as symmetrically as possible. It obviously pays in more than one respect to be a symmetrical man. One dominates others more easily and is altogether healthier.

Asymmetry in the Mind

However beauty appears or how it is created, it must be perceived by the mind. The brain, shown in the following figures, is divided in two (Fig. 6.9) and both its parts are asymmetrical (Fig. 6.10, Table 6.1) from the

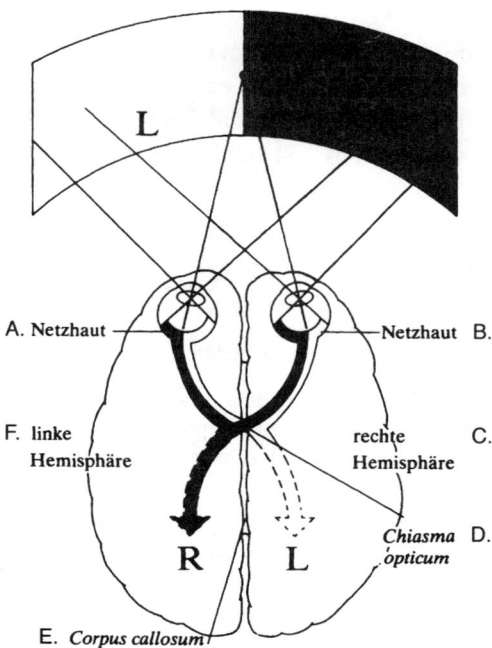

Figure 6.9.   If we focus on a (fixed) point right in front of us, we will see the entire picture in two separate halves of the brain. A. retina; B. retina; C. right hemisphere; D. chiasma opticum; E. corpus callosum; F. left hemisphere.

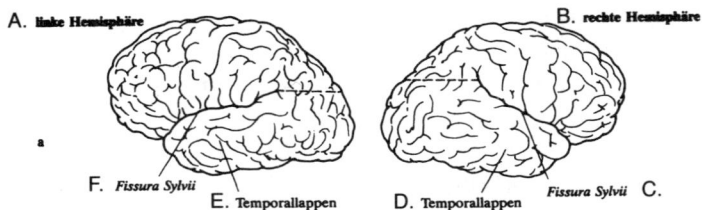

A. linke Hemisphäre   B. rechte Hemisphäre

a

F. *Fissura Sylvii*
E. Temporallappen   D. Temporallappen   *Fissura Sylvii* C.

Figure 6.10.   In the right hemisphere the so-called lateral fissure (Fissura Sylvii) rises higher, and in the left half of the brain it is the temporal plane that takes up the greater area. We assume this area to play a role in the understanding of speech. A. left hemisphere; B. right hemisphere; C. Fissura Sylvii; D. temporal lobes; E. temporal lobes; F. Fissura Sylvii.

outset. The development of asymmetry in our brains sets in before we are born.

Experts speak of the lateralization of the human brain and refer to the fact that it manifests itself differently for different people. Left-handed people and right-handed people can easily be described as having a mirror image exchange of functions. I will not discuss these extremely important individual differences, but will only observe the average case of the average right-handed person whose normal (biologically determined) lateralization has not been altered by a special form of upbringing—in a country

Table 6.1.   Key Words for the Lateralization
of the Brain, Understood as the *Cum Grano Salis*

Left hemisphere	Right hemisphere
Spoken language	Metaphorical meaning of language
Reading and writing	Melody of spoken language
Rhythm of music	Harmony of music
Color naming	Spatial orientation
Verbal thought	Visual/spatial
Sequential	Simultaneous
Digital	Analog
Logical	Holistic
Analytical	Synthetic
Rational	Intuitive

with an unusual language or in a household with unusually abundant musical opportunities. In the brain of such a person, things work quite well asymmetrically (Table 6.1), and we might pose the question: To what extent does this division of the brain functions have anything to do with the classification of the seen (or perceived) in beautiful and not beautiful?

In an experiment that goes back to the American psychologist S. Bresken, we can show people geometrical figures in such a way that they see them either with both halves of the brain or only with one hemisphere. The figures can afterward be distinguished by whether they meant something to the test subject, such as whether they had a simple form and somehow seemed complete. In technical jargon the term is *pragnanz* (Fig. 6.11). This expression derives from gestalt psychology and refers to images that have a meaning in and of themselves. Thus a test subject can be asked which figures are preferred, the drawings with pragnanz or their counterparts.

In this experiment, we discovered that figures with pragnanz are preferred most when they are perceived or seen with the right hemisphere. We discovered, on the other hand, that men generally act as a homogeneous group whereas women are more individual in their choices.

If we stand before a picture in a museum, we can't help but take in what is actually seen on the left, on the right (left) half of our brain. If a picture is extremely asymmetrical, then the focal point of attention has to be either more toward the right or the left; it may be that our average right-handed person's judgment of the image—beautiful or not—is dependent on this position. Once again, science finds something to measure, however without surprising results. Science already discovered a long time ago that average observers prefer an image whose focal point lies more to the right; thus they prefer the one seen with the left hemisphere.

Figure 6.11.   Bresken figures that are supplied with or without pragnanz; "x" marks the figures with pragnanz.

When the test subjects are asked to explain what they find better about the picture that favors the right, they answer interestingly enough that the whole scene creates a more balanced impression. People spontaneously (naturally) tend to perceive images in the right hemisphere, which apparently leads to directing their attention automatically toward the left. This effect can be proven experimentally if a test subject is shown images that depict almost perfect mirror image symmetry and is asked whether these pictures are symmetrical or appear to favor the left or the right. The answer of the right-handed people is that the picture favors the left.[10]

To summarize, we can say that versions of the pictures with a right orientation are preferred because they look more balanced and thus more beautiful. The asymmetry of the stimulus rectifies the inherent directional preference. Each perception of an image perceptibly activates the right side of the brain, and when the corresponding tendency toward the left is confronted with a shifting of the focal point to the right, giving the left half of the brain more to do, then both hemispheres can work together harmoniously and produce this impression on the brain's owner.

A. Lichtrichtung in über 2000 zufällig ausgewählten Gemälden

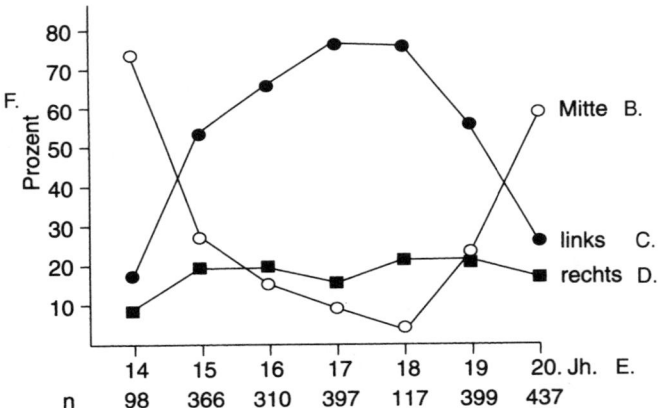

Figure 6.12.   Distribution of light direction in over 2000 paintings of European art, as determined by O. J. Grüsser. A. Light direction in over 2000 randomly selected paintings; B. middle; C. left; D. right; E. -th century A.D.; F. percent.

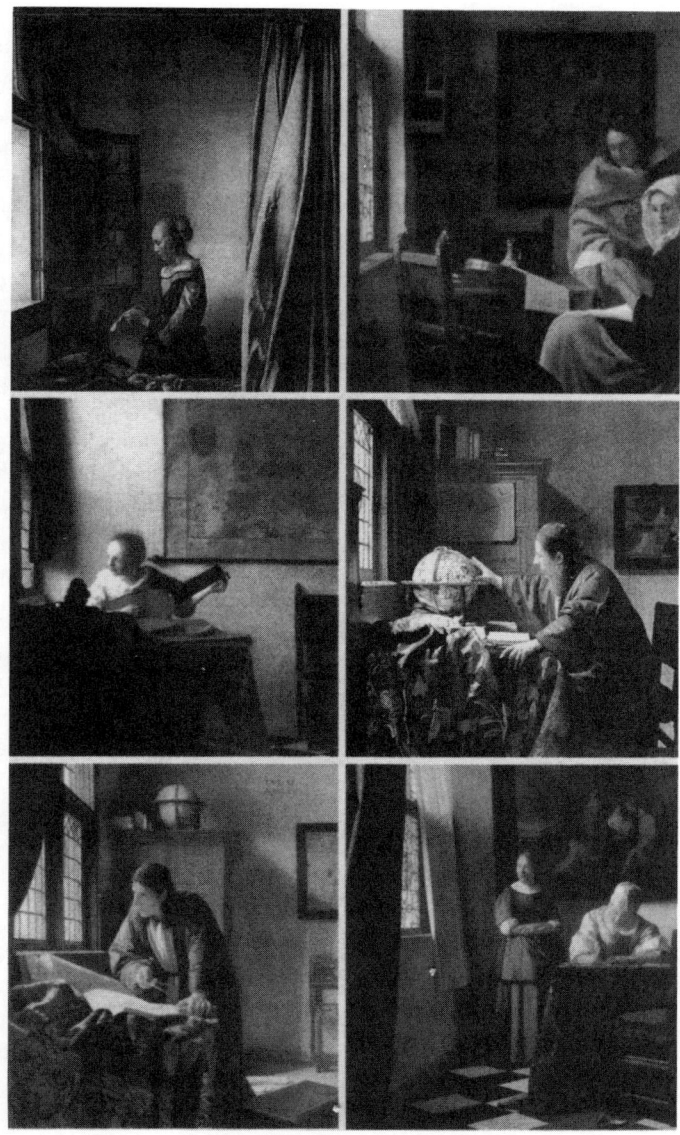

Figure 6.13. Six of Vermeer's famous paintings. Light never falls from the right in his interiors.

Asymmetry of Paintings

The question as to whether right–left asymmetry has an aesthetic component was probably posed for the first time by the great Swiss art historian Heinrich Wölfflin who, in his last book in 1941, reflected on "right and left in paintings." He suggested that "the observer begin at the lower left side" of the painting and "proceed to the right side, where the more important content," at least in paintings of the Renaissance and Baroque, is displayed.[11] Today we can pursue Wölfflin's presumed left–right eye movement experimentally. As a matter of fact, we have discovered that when we look at landscape paintings, we begin on the left and proceed with the eyes to the opposite side.

The question we might ask now is whether or not painters have conformed the composition of their paintings to such asymmetrical requisites of perception to which even they are subject. One neurobiologist, Otto-Joachim Grüsser, discovered a conspicuous asymmetry in art and remarked that in most paintings considered Western works of art, especially from around the time of the Scientific Revolution, the light falls from the left[12] (Fig. 6.12). In this context the paintings of Vermeer, in which the intensity of light consistently comes from the left, are especially convincing (Fig. 6.13).

Grüsser was tempted to "speculate whether this left–right asymmetry in light direction found in paintings of Western art could be correlated with the elementary left–right asymmetry of light distribution" that we all experience when as newborns we lie in the left (nearer the heart) arm of our mothers, or whether it has something to do "with the general left–right asymmetry in space perception caused by the fact that the right hemisphere is dominant in spatial tasks."[13]

Grüsser, following in the footsteps of Erasmus Darwin, knew that the art historian is not interested in the right–left division when paintings show many more complex and subtle shapes and spatial divisions. For a neurobiologist, however, who in terms of science is looking for possible elementary rules governed by aesthetic perception, these observations are anything but boring. If we are to understand the connection between the brain and the culture of certain people, we must start small—with a comparison of the asymmetries that we like in pictures to the asymmetries in our minds that influence our perception.

7   *Perception and Selection: The Origin of Beauty and How It Determines Our Lives*

> *By beauty I mean, that quality or those qualities in bodies by which they cause love, or some passion similar to it.*[1]
>
> Edmund Burke

The purpose of evolution, according to Joseph Brodsky in his essay entitled "An Immodest Proposal," is neither the survival of the fittest nor of the strongest—because then all men would look like John D. Rockefeller or Arnold Schwarzenegger.[2] The goal of evolution could also not be the survival of the so-called least fit or strong, for example, "the defeatist," for then we would all look something like Woody Allen. The purpose of evolution must be something else, and the Russian–American poet had the courage, shortly before his death, to commit himself in essay form to an "immodest proposal." For Brodsky, it's beauty, for as he wrote, "The purpose of evolution, believe it or not, is beauty." For it is beauty that "survives it all and generates truth simply by a fusion of the mental and the sensual."[2]

These are great words that, in my opinion, deserve consideration, for they would merge—as Kepler did with his archetypal notions—a (mental) interior and a (sensorially grasped) exterior to make cognition aesthetically possible.

Whoever earns his or her living in the operations or observations of science might consider Brodsky's bravadoesque statement (if it reaches their ears at all) as pure nonsense. Scientists do not speak of a purpose of evolution in the context of biology. This is part of the basic assumption under which modern science has operated very successfully since its first appearance at the beginning of the seventeenth century—that in general

135

the observation of natural processes as goal oriented is sterile, as "a virgin devoted to God is barren."

Or so it was according to Francis Bacon in 1623 when the English philosopher gave the famous first signal that required science to be pursued in the service of the material well-being of people and to relieve the certainly very difficult existence of our ancestors at that time. According to Bacon, it was pure and simple: Nature, as opposed to an individual human being, has no purpose, and therefore, if we want to understand it, we shouldn't even start looking for one. We should concern ourselves instead with discovering the laws of nature. If we know them, and direct them, we can use them to overpower the forces of nature.

Knowledge Is Power

Man must first heed nature to master it, according to Bacon, and through his enlightened idea (still practiced in our time) something decidedly new came into the world: the human being who subjugates what confronts him in nature. The subjugated is called in Latin *subiecto*, and when acting submissively, the Latin adverb expresses this as *subiective*. At the beginning of the seventeenth century, when Bacon denied science every purpose, he gave it—in return—its object, that which stands in contrast to the subject, called in the language of Cicero the *obiectus*. In the course of the Scientific Revolution of the seventeenth century, the world was suddenly split into subject and object, whereby only subjects—people—could pursue a goal, but the object could no longer do so; now the object, nature for example, could only be mastered.

Posterity has taken Bacon's motto "Knowledge is Power" to heart and kept it in mind. Thereby the persuasive and successful thought has arisen that in science a piece of knowledge turns out to be true when we can use it. Giovanni Battista Vico expressed this thought in the early eighteenth century with the Latin sentence, "Verum et factum convertuntur," which, when translated, suggests that the *deed* (factum) is a criterion for the *truth* (verum). Feasibility influences accuracy.

This idea is found in an earlier form in Galileo's work. He demanded that science should only concern itself with what can be measured, and accordingly it should make everything worth knowing measurable. This demand for complete quantification goes together almost self-evidently with the abandonment of sensory perception as Galileo stated clearly in his "Il Saggioatore":

I am of the understanding that nothing more is necessary to generate sensory perceptions in us such as taste, smell, or hearing of the external world than the availability of form, figure, and slow or quick movements. I believe that if ears, tongues and noses were taken away, forms, figures, and movements would remain, but not smell, taste or hearing. When the latter are taken from living creatures they are nothing more than names.

Beautiful Animals

It's no wonder that in a world aligned exclusively with the practical and the achievable and in which science operates accordingly, we would develop the idea that evolution follows suit: Nature, by means of natural selection, tries to generate the strongest and the most powerful who can hold their own and drive the others away. "Survival of the fittest" is the well-known phrase, according to which only the fittest individuals of a species survive in the long run, and they do so by bearing and successfully raising many young.

Now and then we still hear the accusation that this formulation is a tautology, for success in reproduction is ultimately defined by fitness. To say that to reproduce is to survive is a banality (the survival of the survivor). If in the meantime we have long understood that this criticism misses the point, we continue to overlook that in the course of evolution certain qualities have developed (survived) besides those that serve in the fight for survival. I am not referring only to the frequent symbioses where nature skillfully cooperates. I am referring here to the qualities of luxury that might be found in all their glory in birds, for example, the bright colors of pheasants, the magnificent tail of a peacock, and the plumage of the bird of paradise (Fig. 7.1). Many living creatures are equipped not only with practical qualities. They are also simply beautiful.

When the word *beautiful* is used for the bird of paradise, it obviously means only that I, the observer, find it beautiful. When we talk about the behavior of animals, and everything that happens as it should in terms of science, beauty and the perception of beauty cannot be discussed casually. In this case, *beautiful* is used to express the attractiveness of an animal that creates an advantage for it in the search for a partner, meaning that it makes reproduction possible (in the time available). That the peahen generally sees her peacock or his fan from behind, from what I observe, I mention here only peripherally.

Figure 7.1. The king bird of paradise that—without counting the center tail feathers—is about 15 cm long.

The Female Choice

As I said before, we have long been aware that in the course of evolution, besides practical qualities, creations have also emerged that—from my viewpoint—qualify as beautiful. In this context we can mention that, last but not least, the human form itself has a claim to beauty (even when the concept of "beautiful people" has become somewhat derogatory). As pure survival machines, the members of the species *Homo sapiens* would likely have a completely different appearance than they do. I don't—as most wouldn't—consider every body a beautiful one, but the remark by Socrates that the human form is more beautiful than that of an animal begins to make sense in evolutionary guise.

Evolution has generated not only fit creatures but also beautiful ones. To understand this dual development, we can try to find a single mechanism for each of the two qualities, and if possible, ones that work together toward the same goal. Its association with the athletic idea of fitness exists in the well-known and accepted explanation of natural selection, to which I will now turn as briefly as possible.

Since the days of Charles Darwin, scientists have understood and explained the adaptation of species to their environments by a connection of several observations. Apart from the fact that all species could produce sufficient offspring, organisms appear in each generation that deviate slightly from one another, and are distinguished by arbitrary genetic changes (mutations). Many variations are hereditary as a result, and these innovations are handed down in abundance to the next generation, which puts their carriers in a better position to find their way in their respective environments (niches). The ones that have adapted best to the climate and to the nutritional supply and are able to defend themselves in the face of natural enemies in the given area will produce the most offspring and to this extent determine the development of their own species.

Natural selection is only concerned with external criteria. It determines the survival chances of a whole species; the beak form of the finch, the thumb of a panda, the neck and tail of a giraffe, the flipper form of fish, or the sonic radar of a bat can be explained in the same way.

All of this has long since been a secure stock of biological knowledge. Yet how do we see the reverse-striped feathers of the wood grouse (Fig. 7.2)? What, for example, is the red spot around its eyes meant to lure? And what does the red-tufted malachite sunbird get for the scarlet-red crest on its brilliant green feathers?

Figure 7.2. A wood grouse from the wood grouse family; the dimensions range
between 70 and 80 cm (from the *Brockhaus Encyclopedia*, Vol. 18, Leipzig: F. A.
Brockhaus).

Of course, someone who perceived nature and its creations as sensi-
tively as Darwin did couldn't help notice these luxurious oddities among
animals, and accordingly he offered an evolutionary explanation for them
(which was then just as predictably overlooked for a long time). Darwin
spoke of the "female choice" with which females look for their partners in
certain species. In their attempt to be chosen and thus win, males will stop
at nothing to be especially gloriously endowed. By far the most well-
known example is the sage grouse that lives in the western United States
and holds the entire performance of partner choice in a kind of mar-
ketplace. All of the males take their place in a line and present themselves
by inflating the air sacs in their feathers for show. While they ruffle their

feathers, the females stroll around the males. They devote a long time to the test of perception before they decide to let their choice be known by crouching down in front of the accepted male.

Darwin could propose the female choice because in fact—with peacocks and pheasants—the males are the ones that develop in a particularly beautiful way, and they develop beautifully solely to achieve their purpose of impregnating a female. (Here is the uniformity in the variety of nature, which in spite of all repetition remains beautiful.)

The asymmetry of conception—the fact that there is a female choice but not a male choice—Darwin explained with the observation he had made quite some time before that males are prepared to mate with every female and often do not make a distinction between females of one or another species. Females, on the other hand, have valid reasons to be on the lookout for the suitable male. Darwin discussed this point in more detail and referred to the reason for this crass difference as relating to the principle of investment. He argued that because a male has enough semen to impregnate many females, his investment in a single copulation is minimal. A female, on the other hand, produces relatively few eggs and invests a lot of time and energy in hatching the eggs, carrying the embryo to term, and taking care of the brood.

This argument, declaimed unceremoniously in modern form as "parental investment," is demonstrated especially forcefully when a hen can take care of the brood all by herself. She can do this easily when—in the case of the birds focused on so far—the young develop quickly (called nidifugous birds). It is only with their counterpart, the nidiculous bird, that the mother needs the help of the male in raising the young, and in so doing the females lose at least part of their freedom of female choice.

As a leading example, I start with the bird of paradise whose offspring leave the nest early, whereas the crow, for example, begins its life as a nidiculous bird. There couldn't be more external difference between the closely related birds: Gray-black crows stand in contrast to the gloriously decorated bird of paradise. The differences can easily lead one to the same conclusion Darwin drew: that female choice is evolution's way of generating beauty.

Beautiful Sexual Selection

The principle of the female choice must be understood as a secondary mechanism of the evolutionary event, and we then begin to speak of sexual selection and put natural selection to the side. The comprehensive external selection—to draw a simple portrait of evolution—has taken precedence in the ancestral history and makes possible an optimal adaptation to the prevailing conditions of life that are climatically and geographically determined and are farther from resources as well as from living creatures that are trying to get into the targeted niche.

When adaptation to the environment is no longer an issue, sexual selection becomes even more important, and it can also work this way for the male, when—as is the case with hooved animals—males claim several females as sexual partners and rivalry ensues. The more forceful specimens survive and the next generation consist of many "born rivals." When sexual selection leads to rivalry, nature produces the battle-ready males that are endowed with courage and endurance.

As important as these qualities are, they only refine or expand the basic principle already modeled by natural selection. On the other hand, when sexual selection means female choice, evolution takes a strategy that is internally rather than externally oriented. In the words of Hilde Neunhöffer, who has presented a comprehensive analysis of the meaning of female choice in human evolution and who has done so outside the traditional channels of science, "it is no longer strength, ability to hold one's own, weaponry, adaptability that are the deciding factors in the further development of a species, but color, beauty, and charm."[3]

That aesthetic points of view are in charge of steering the evolution of a human being, and that sexual selection has given up its role on the sidelines and shifted to the center of the explanation, has been determined in a recently published history of sexuality. Its author, the British science journalist Matt Ridley, wrote:

> The evidence is beginning to accumulate that humanity is a highly sexually selected species and that this explains the great variations between races in hairiness, nose length, hair length, hair curliness, beards, eye color—variations that plainly have little to do with climate or any other physical factor. In the common pheasant, every one of the forty-six isolated wild populations in central Asia has a different combination of male plumage ornaments: white collars, green heads, blue rumps, orange breasts. Likewise, in mankind, sexual selection is at work.[4]

The female choice leads to charm and beauty and the problem consists in finding a reason why human ancestors have granted women this opportunity. How did women get the chance to choose? Perhaps by guarding the fire, women earned a decisive influence and perhaps by bringing children into the world, they strengthened their position, whereas the male population in early human societies probably had very little understanding of their own role in contributing to the generation of offspring.

Good Genes or Just Good Taste?

Before I investigate further how the female choice has made human beauty possible, and before I analyze any further which behavioral tendencies we have as a result of sexual selection, I need to turn our attention again to the animal kingdom to clarify the question of how this internalized choice actually operates. Why do females look for the most beautiful male? Why do they develop, as Darwin formulated, an aesthetic sense for male ornamentation? And what do the males want by showing off their ornamentation? What is the biological basis for sexual selection?

The countless explanations offered by scientific literature can be divided into two groups. In one we find notions that can be summarized as "the theory of good taste." According to this theory, the peahen, for example, searches out the most beautiful peacock simply because she is interested in passing on his beauty to her sons, and so they in turn can attract as many possible females, and so on.

Beauty is here more or less a virtue in itself, and as attractive as I find the thought, it seems capricious and leaves to pure chance that at some earlier time a hen chose the most beautiful rooster. She might also have chosen the largest or made her choice according to some third criterion.

It makes sense that females have enough reason to prefer long peacock tails if enough other females do. It's not clear, however, in the context of these models just what makes long feathers beautiful in the first place. In fact, long feathers are more of a hindrance to what we would normally call the fight for survival. We might say that the rooster that has persisted despite his decorative plumage must be especially capable and thus better suited to survival than the others. Yet hairsplitting like this is pointless in terms of science.

At this juncture it is worth looking back to Darwin. To paraphrase, in the search for a partner, males are interested in quantity and females are interested in quality. The females cannot mate as often as they would like, at least not in terms of reproduction, because they are the ones that carry the young and take care of them after they are born. Because they cannot raise many children in the course of their lives, females have to be much more choosy, and this means that to get only the best genetic material for their offspring, they must look for the best partners. In biological terms, to be successful, a female has to find a means of recognizing the partner with the best genes. Thus she must view the external form. Females recognize the quality of the gene by the beauty of the male.

We can't conclude from this that good genes lead directly to beauty (or that there is such a thing as a gene for beauty). However, we can say that outward beauty reveals good genes within. In this conceptualization, beauty is not a virtue in and of itself. It has a purpose: to point to the intrinsic worth of the genetic material so a female, in the case of reproduction, will conceive and mix in her own genes.

According to this theory, peahens wouldn't be on the lookout for long tail feathers out of some kind of inborn aesthetic sensibility for the way a suitor strikes up his beautiful fan (Fig. 7.3); peahens would be studying that fan much more closely—meaning taking in its color and pattern—because they would be able to recognize the quality of the gene, or to be exact, the quality of the genomes this way. Genomes are understood as all the genes together. The trait that concerns hens most is the susceptibility of this male genome to parasitical infestation.

Parasitical infestation in the context of sexuality and beauty must be a little disturbing at first. But in the larger context of my scientific picture—the idea of evolution—it has everything to do with an enduring contest between different forms of life—the duel between parasites and their hosts.

One of the fundamental convictions of modern evolutionary biologists is that the central purpose of sexual reproduction—the reason why evolution has developed this complicated way of passing down genes—resides in making one's own offspring as resistant as possible to all omnipresent parasites. In other words, by choosing the most beautiful male, the female can capture the genome that is least susceptible to parasites. If we take another look, we can see how she does it. Spreading out a giant peacock fan is not an easy thing, and a female onlooker must be able to spot deviations in symmetry almost immediately. The degree of symmetry in the pattern becomes a direct measure for the ability of the peacock genome to fulfill its task of being perceptibly (!) resistant to parasites.

This kind of theory can only find acceptance when it stands on experimentally solid ground, and in the meantime it does. Detailed investigations have revealed that, in fact, this species of colored birds is infected

Figure 7.3. The peacock (male) can reach 2 m long including tail feathers (from the *Brockhaus Encyclopedia*, Vol. 17, Leipzig: F. A. Brockhaus).

with a high concentration of blood parasites. We can even say regarding them, as for many species of fish, that the more parasites there are, the flashier the species will be.

This subject leads to an altogether strange connection that fascinates people in the upper levels of the intelligentsia. I am referring to the connection between disease and beauty, a common theme in literature that has its point of departure here.

Beautiful Symmetries

Sexual selection leads to what sexuality generally leads to, ensuring healthy children or to put it bluntly, ensuring that offspring are as resistant as possible to parasitical agents. We can see how this is achieved by viewing the symmetry of characteristics or of body parts, for example, the tails of swallows. The tail's symmetry is the deciding factor for success in choosing a partner, as confirmed in many investigations and observations.

Again and again we discover that the symmetry of organisms says something about their fitness. In scorpionflies, for example, there is a clear correlation between the symmetry of their wings, in respect to size and form, and their success in hunting for prey, as well as their ability to defend their seized victims (Fig. 7.4).

In any case, a researcher can certainly verify and measure the balance of the wings of the scorpionfly, but when it comes to matchmaking, the female has no chance at all to take a look at the wings to recognize their symmetry. Presumably there are other external means—aroma, sound—that are called into service to apprise her of the male's state of health. As humans able only to see the wing symmetry, we regard this one visible sign of health solely for its beauty.[5]

With humans themselves, if I may make this giant leap in one sentence and without forewarning, we can say unambiguously that a symmetrical face is a sure sign of good health. At any rate, an experiment in American universities in 1996 in which about 100 students kept extensive diaries on their physical condition revealed that people with symmetrical facial expressions suffered less from insomnia, stuffy noses, general irritability, and envy.[6]

Symmetry in the human face is right out in the open, but it is not the deciding impression responsible for whether we pause in awe of another person's appearance. The deciding impression is actually created by certain

Figure 7.4. Scorpionflies are around 20 mm long and belong to the order of scorpionflies.

deviations in the symmetry of the average face. The biological attractiveness of an average face can be explained by the simple consideration that individuals equipped with average characteristics of a population carry fewer potentially genetic mutations.

That special face (Fig. 7.5), as empirical studies in science show again and again in the United States, Great Britain, and Japan, is produced in women by a higher forehead, fuller lips, and a smaller chin. The preference for a very tiny jaw in women is interesting because it is exactly the opposite for men, where a strong chin makes a handsome face.

It is mainly the lower portion of the face that develops differently in the genders after puberty and in this way reveals outwardly the inner influence of specific hormones. A small chin for a woman may have been a sign that betrayed to her suitors her potential for pregnancy because the hormone estrogen supplies this feature. In men, androgen creates a big

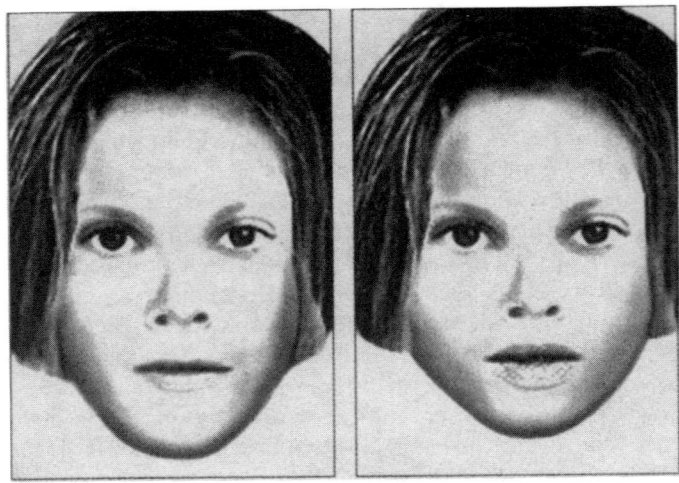

Figure 7.5. Two of the computer-generated faces; the left corresponds to the average face and the right unites all of the features considered ideal by most people.

chin. Evolutionary biologists sometimes call this "truth in advertising" because it is a true sign of higher resistance to disease.[7]

Beautiful Bodies

Our faces may be nature's business cards that we hand out whenever we make contact, but the body, the physique, is the real issue when the argument concerns sex appeal and biology. If we can imagine our early ancestors unable to cover themselves with clothes, it becomes easier to see how the shape of the body might excite more interest than the subtleties of the face.

The physique—at least in contemporary life—primarily captures men's interest. If we think about the female figure and how, in the course of a lifetime, it undergoes measurable changes, men's interest in it becomes more significant. In the evolutionary context, the connection between appearance and fertility has long been supposed because a woman's ability to bear young is easily measured by her beauty.

Before puberty and after menopause women have the same waist measurements as men. During puberty, however, when boys develop muscles and their bones grow, girls begin collecting the reproductive fatty tissue and storing it in their hips and thighs. These pounds carry the calories needed for the nutrition of the fetus during pregnancy. Just about everything that influences a woman's figure is related to fertility. Healthy, fertile women are recognized by a ratio between waist and hips that lies between 60% and 80%. Their waists are between 60% and 80% of the size of their hips, apart from total body weight.

Women can of course bring children into the world even if their waist–hip ratio lies somewhere outside of this range, but as a Dutch study has shown, even a slight deviation can introduce complications. With 500 women who applied for *in vitro* fertilization, the probability of success went down when the waist–hip ratio varied by 10%. From an evolutionary standpoint, it is unimaginable that men would not react in some corresponding way, and so it is no mystery why men classified female figures with a 70% waist–hip size quotient as beautiful, as confirmed by a simple experiment (Fig. 7.6). Male test participants between the ages of 8 and 80 decided with an overwhelming majority in favor of the silhouette designated as N7. She is in the range of normal weight, and in fact in the middle range in which the ratio between both body measurements indicates health and fertility.

The Vision of Perception

The male gaze at the female figure is a far cry from the "female choice," but both cases do involve the decision process connected with beauty, and this decision process is closely connected to the prerequisite ability for taking in and evaluating all these signals—perception.

When we really devote ourselves to these questions, we start to think that, as human beings, our sense of aesthetics must follow some kind of general principle of perception developed during the evolution of biological signals. The preference for symmetry does not necessarily need to arise as an indication of the quality of the signal giver. Its high rate of preference might have simply been a by-product of the solution to the problem of recognizing objects or bodies independent of their position and their orientation in a visual field. The existence of a sensory inclination for symmetry might have been—I am always reading this—something in the course of our ancestral history that evolution and natural selection took advantage of while biological signals were developing. In the course of

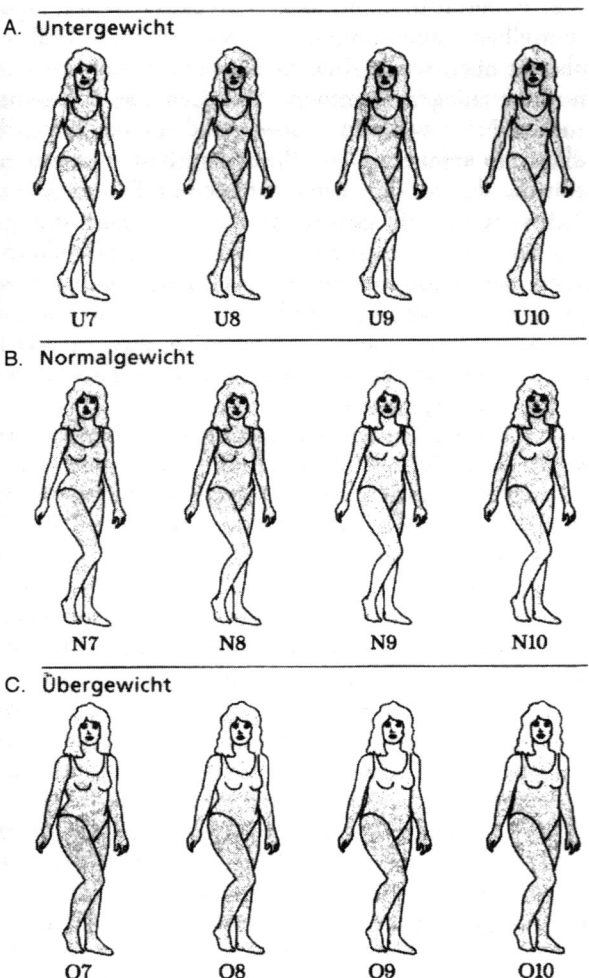

Figure 7.6. When male observers were asked to rate the depicted female figures with varying hip and waist measurements, most participants chose N7, N8 and U7 were next, and O10 came in last. A. Underweight; B. Normal weight; C. Overweight.

cultural history, this predilection for symmetry might have been used by artists to heighten interest in their pictures.

It's interesting that many signals that animals use for communication humans classify as beautiful—the color and pattern of flowers and butterflies, for example. It is striking because the signals arise for organisms that in the course of evolution have become equipped with visual systems and mechanisms that function according to completely different principles.

One problem confronting animals (and humans) is recognizing living creatures and objects in various positions and in differing orientations within the field of vision. The retina focuses on a type of image of an object that can be seen from a certain position. The object in view is geometrically transformed during this process. The work of perception researchers now consists of the idea that the necessity to generalize many such transformations can lead to preferences for symmetries and to the evolution of symmetrical signals.

The answer to the question of why nature loves symmetry might be found in a "conspiracy" of geometry and neural architecture that gives preference to symmetrical patterns over asymmetrical ones. In fact, scientists have shown that artificial neural networks in the position to recognize images tend to be interested in symmetrical patterns.

Multiple networks that have minimal differences are confronted with images (patterns), the tail of a bird for instance, that they are supposed to either recognize or overlook. The best network is chosen, the others are discarded, and just as in evolution the discarded leave behind no "offspring." The surviving network now serves as a model for new, arbitrarily varied networks, as in genetic mutation. This cycle is repeated many times until a network is generated that has the ability to recognize patterns.

Attempts show that the "trained" networks react better and better to symmetrical variations among the images provided. It comes down to a coevolution between signal and receiver and ultimately to an overwhelming preference for patterns that are symmetrical in rotation (Fig. 7.7).

The Perceiving Human Being

As difficult and interesting as such investigations may be, they do not lead to a more precise understanding of the role that perception plays in the female choice. Symmetry-loving neural networks are all well and good, but human beings must achieve more with their perception to get by.

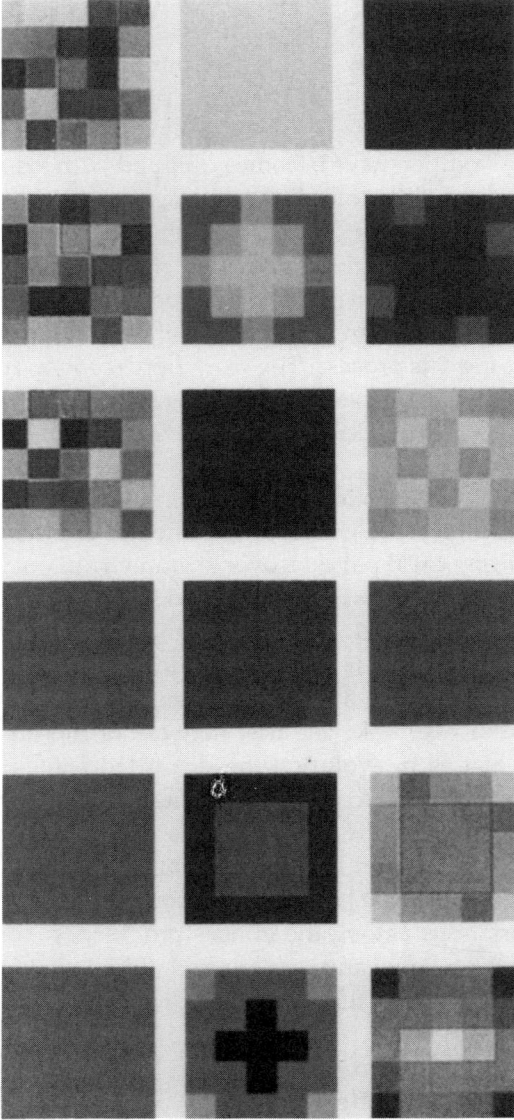

Figure 7.7.   The signals used in the experiment consisted of the patterns as shown. In the left column are the patterns that were ignored; in the right, the less effective patterns. The signals were effective when they were symmetrical. The center column shows the patterns that emerged during coevolution (from Enquist, M. and Arak, A. *Nature*, 372, 1994, p. 170).

The question worth answering is whether we can find some evidence in scientific quarters that Brodsky was right and that evolution stops at nothing to generate beauty.

Natural beauty is easy to understand to the extent that we can say that biologically relevant signals are also aesthetically influential. The other way around wouldn't add up to much and would only be taking a single evolutionary step. Beauty of the artificial, artful kind is much more complicated to understand, and when it is understood at all then it must—in this context—have something to do with the choice of partner and with the tricks that people use to increase their advantages.

In the course of our discussion, we have been assuming that human beings are no longer subject to natural selection, but rather to sexual selection. The presence of other men and women is evidence enough for this transition, which must have happened 100,000 years ago. Many scientists believe that ultimately "only humans themselves could provide the necessary challenge to explain their own evolution," as expressed by zoologist Richard Alexander.[8]

It was about 100,000 years ago when for the first time a development did not lead to the adaptation of the surrounding nature, but to the adaptation of human values and perceptions. Human beings themselves now seek out each new advantage. A man might have the experience that women no longer prefer the strongest applicants but rather the most humane, the ones who are considerate and have developed the most affection and sensitivity.

At the very beginning of the path to *Homo sapiens*, women had to learn to perceive others, for example, their children, so they could recognize their interests. This state of affairs may explain women's ability to see through their fellow human beings better than most men. Women are responsible for recognizing their partner's thoughts. Their intuitive psychology far exceeds clinical psychology for farsightedness and for accuracy.

No development of this kind proceeds without a counterreaction, and for obvious reasons those who are "seen through" have looked for ways of deceiving their perceptive partners. It is well known that betrayal and the production of "hot air" play a large role in human communication, but I am going to overlook the negative aspects of human development and lead to a positive aspect of coevolution because I prefer the good connected with beauty over the bad.

Another possibility exists for a man to react to a woman's superior perceptive skills; a man can try to exhaust a woman's ability to perceive by presenting her with something new, for example, a piece of art. In the words of American biologist Geoffrey Miller:

> I suggest that the neocortex is not primarily or exclusively a device for toolmaking, bipedal walking, fireusing, warfare, hunting, gathering, or avoiding savanna predators. None of these postulated functions alone can explain its explosive development in our lineage and not in other closely related species. . . . The neocortex is largely a courtship device to attract and retain sexual mates: Its specific evolutionary function is to stimulate and entertain other people, and to assess the stimulation attempts of others. . . . Just as the peahen is satisfied with nothing less than a visually brilliant display of peacock plumage, I postulate that hominid males and females became satisfied with nothing less than psychologically brilliant, fascinating, articulate, entertaining companions.[9]

If women began to prefer intelligent or creative men, for example, then the resulting creative children would ensure even more creative men, and so the tendency would increase exponentially. The resulting brain would then become as large as the peacock tail is beautiful. This particular development raises the question why women themselves have managed it—to get bigger brains—though this is denied the plain peahens—to acquire feathers just as beautiful as the ones they favor. Clearly, however, perception requires a large brain when the one being perceived starts to deceive and to put on an act, especially when time goes by and his act improves.

When men had to become smarter just to increase their odds in terms of the female choice, it is no coincidence that when the first works of art emerged in the history of humankind, they were exclusively images of women, not necessarily women with ideal measurements, but ones whose measurements accentuated their signs of fertility—take for example the Venus of Willendorf (Fig. 7.8).* The conclusion seems plausible, that it was men who created these first art objects to please women, the ones equipped with the actual power, the power of choosing a partner.

The Given Form of the Beautiful Object

With the emergence of these first art artifacts perhaps offered to women about 30,000 years ago, the feeling began to set in that not only

---

*The empirical basis for this statement is thin and can at any time be disproved by counterevidence.

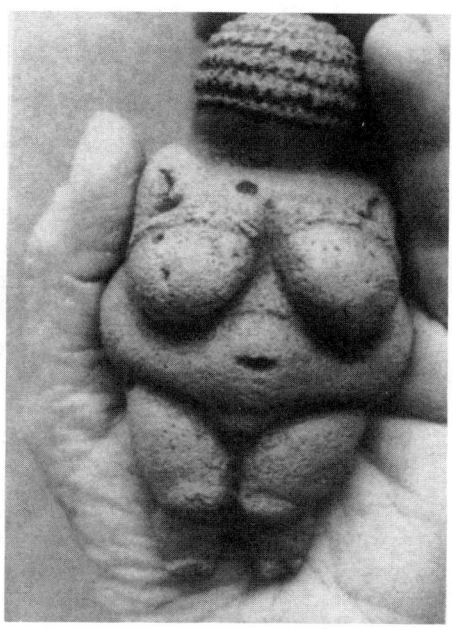

Figure 7.8.   "Venus von Willendorf" discovered in Lower Austria in 1908 fits in the palm of the hand and is a 4-inch chalk stone statuette.

beauty was recognizable in the human form but that there were further qualities to perceive. Women were most likely the ones to have developed a certain feeling for the worth of other people. We develop this kind of sensitivity when we turn our attention toward other people and recognize what makes them special. In other words, we begin to be human. We become interested in beauty and out of this biological motive we make progress from *Homo sapiens,* to *Homo humanus,* and finally to *Homo aestheticus,* the perceptive and feeling human being.[10]

The original concept of beauty, in this biological view, is found in the concept of the beautiful human being, which Goethe considered the highest aim of unfolding nature. For a philosopher of the twentieth century this concept is at best "vulgar currency" as Nicolai Hartmann expressed it, who does not posit his aesthetic over the head of the normally gifted reader. For Hartmann, perception is the "given form of the beautiful object." In his opinion, beauty is directly perceived, and the feeling becomes affectively enriched as it approaches the idea of evolution: "Emotional components are not imprinted secondarily on perception, but are the original components, and objective perception came out of this area relatively late." Emotional reactions adhere "directly and completely to perception." They touch—according to the philosopher—on "the mystery of the first impression."[11]

The first impression, love at first sight, always goes beyond individual parts and takes in the complex whole, which I do not grasp as separate from me even as I am confronted by it. Our first perception only establishes a relationship, a harmony. At first glance the world is always beautiful. Thus most of human experience is aesthetic, as we read in Joseph Brodsky, who started out this chapter. In a lecture in which he reconciled the state, the language, and the poet, the fact that most of human experience is aesthetic means "aesthetics is the mother of ethics," because, he illustrates, a small child crying before a total stranger who draws away or, in contrast, reaches up an arm, instinctively chooses according to aesthetic categories, as opposed to moral ones.[12] Why can't this be true all our lives?

8   New Values for Science: An Appeal for the Aesthetic Transformation of Scientific Research

It concerns the development of the aesthetic function. For the suppression of this is precisely what blinds the occident to the myriad ways that human beings create.[1]

Adolph Portmann

It's strange: The sciences are equipped with the best opportunities for understanding beauty in evolutionary nature and for explaining its attractive efficiency. They also have countless reasons for seizing beauty in the intellectual sphere and for putting its creative efficiency to use. And yet, these days the idea only occurs to a few to call the products of science beautiful and to look for beauty in this region of the intellectual world, although not too long ago we were still gripped by the scientific view of the world or at least seemed enthusiastic about it. In addition to the far-reaching theories on the inner workings of physical matter and of the cosmos, we could be astonished by the whole idea of biological evolution. And although we tried for a while to draw near to this scientific sketch and its truth, many people would rather turn away from it when they observe the scientific world and see, for example, photographs of manipulated and cloned life forms. They stand by helplessly and no longer see a worthwhile goal in science. Earlier, we stood in awe before Einstein's theory of relativity; today people pity a geneticist's fly with 14 eyes or similar creatures that are hard to look at and violate common sense. Earlier we looked to science for progress and then to increase our longed-for power over nature. Today we fear the progress of science and its consequences that seem to increase our feelings of powerlessness.

157

Value-free or Valuable?

Something has gone wrong with science, at least from the point of view of the interested public taking in the unceasing flow of novelties from research centers and institutions with a strange joylessness. The feeling is widespread that somehow something in science has to change. Ultimately, everyone knows that without science, at least in developed areas, survival would not be possible and there would be no future worth having. In this chapter I attempt to give some indication of the direction scientific procedures should take if science as a whole is to move forward. It will be a surprise to no one if I refer to aesthetic elements and suggest a kind of return to the use of the senses, understood as a trust in the perceptible sensory experience.

To put it concisely, I suggest that science should no longer, as it has for centuries, throw all of its efforts into trying to make something useful, but instead that it make better use of its abilities by trying to make or discover something valuable. Because research has not yet tried to make or discover anything that has no use, correspondingly in the future, science should never try to make or to discover anything that has no value.[2]

For many reasons, it remains extremely difficult within the operation of science to find out what is valuable. For a long time, researchers have justified much of what they have done by pointing out that science and its activities are value-free. In other words, they disregard the results of their research and thought regarding considering its value, and they are satisfied with this.

One of the worst outcomes of philosophical thought is called *rationalism*, which is mainly concerned with the "reasonableness" of things. Rationalism introduced the concept of value-free science into the world. In science, it has helped establish that we do not have to spend too much time asking whether the objects of our desire and the results of our appropriations are ethical violations of value or are acceptable, whether they carry in them healing or harm, or whether they serve conservation or devastation (potentially). Of sole importance is the scientifically precise acquaintance with things and the experiment that will produce something profitable, for example, a form of technology. It is no wonder that most researchers have lost no time in accepting these kinds of requirements, so that they could freely choose their objects and be able to test and develop value-free practical procedures.

The Role of Perception

The construction of the value-free only makes sense as long as it can proceed without perception or as long as we can do without the perception of the world. On the other hand, when sensory experience and dawning awareness come to rights, the world of science immediately begins to look different. Whether researched or not, "things" have a distinctly perceptible value—and we finally notice this when something is recognized for its beauty. Ultimately, it is by perceiving with our senses and recognizing beauty that we come to regard a thing as valuable and worth preserving. It's worth taking a look at this beauty. In other words, scientists of the future must try to understand nature less through concepts and more through the senses to reveal the beauty of science. Unfortunately, even in the scope of today's active environmental protection movement, nature is conserved not because it is beautiful but because it has become damaged and ugly. A convincing relationship with the environment must have an aesthetic source. The value of environmental conservation must be perceived by its beauty.

In these or similar words, the program of an aesthetically oriented science—the program of aesthetics—could be formulated. In the meantime, this program must defend itself against many biases, for ultimately the rational sisters, science and philosophy, continue to maintain that deception goes hand in hand with sensory experience, which is responsible for generating a kind of confusion in the mind. Kant, for one, never tired of refusing the "turmoil of the senses" as an instrument of knowledge (without noticing that at best he means overwhelming stimuli).

However, as we might easily guess, the opposite is more the case, if we would become aware of the many subtle perceptual constancies (Fig. 8.1) such as the color, shape, and dimensional constancies. If we consider constancy as it relates to color, readers can easily perceive that they observe the paper of this page as always appearing white independent of the (natural or artificial) light they are using to read. Physically speaking, utterly different wavelengths of completely different intensities of light reach their eyes according to whether they read by sunlight or by a neon lamp. In their perception, however, the page remains white, and such a color constancy makes sense if we argue from the notion that this is how evolution has generated human beings. Evolution must do this because only with the help of invariant colors like red, green, or blue is it possible to find one's way in the world.

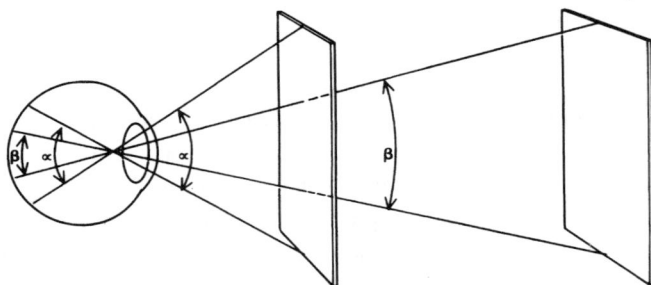

Figure 8.1.   When an object is seen from a great distance, a smaller image appears on the retina of the eye. The perception might suggest to the observer that he is seeing a smaller object. Consciousness, however, is told correctly that the object's dimensions have stayed the same and that it only seems farther away. Besides this dimensional constancy of perception, science also knows color constancy: Colors do not change to any great extent even if the light illuminating them changes. It's the same with shape constancy: Objects keep their shape after a rotation, even if their projection on the retina rotates. The science of physiology has a hard time explaining perceptual constancy.

Yet the rational bias against the unreliable senses and our perception that is susceptible to deception dies out slowly. We still believe that thought alone will help us gain a greater measure of certainty, and so the idea digs its heels in even deeper. Deceptions of the senses do exist (Fig. 8.2)—such as the well-known illusion of the moon, where the moon rising on the horizon will appear bigger than the moon already in the sky, but these phenomena are easier to penetrate than the many larger conceptual deceptions under which Western culture suffers, and that to this day open the door to all kinds of seduction.

It is not so much ideas that carry sensory experiences, as perceptual constancy. Perceptual constancy permits a very special kind of organization of the "turmoil of the senses." It is an arrangement with which everyone is familiar who has seen the "man in the moon" or some kind of familiar face in a cloud in the skies. These cases show the tendency of perception to summarize stimuli to generate a whole or a familiar pattern. Perception is looking—as already mentioned—for a good figure, and it would be great if today we once again took the "perception of form as the source of scientific knowledge" as seriously as Konrad Lorenz did in the 1950s.[3]

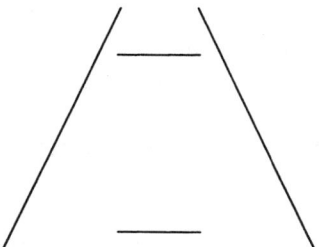

Figure 8.2.   The Ponzo illusion: The upper (perceived as being farther back) line looks longer than the lower line (perceived as in the forefront). This illusion can be explained by dimensional constancy (Fig. 8.1). Perception "thinks" that the upper line is farther away. Because its image is the same size on the retina as the image on the lower line, perception tells the brain that the upper line is longer in reality. The interesting part of this kind of sensory deception is that the illusion doesn't go away when we look at it with full awareness of the concept. What's more important in this context is that it is constancy that doesn't go away.

Perception has a function that beyond its contribution to knowledge leads directly to moral and ethical questions.[4] Before I discuss this aspect, I would like to address another subject that will help make the significance of such an aesthetic transformation visible and desirable.

The Pleasure of Thinking

If we give some thought today to natural science, or even spend some time contemplating it, we soon realize that the subject of discussion is still centered on the truth of its theorems and particularly on the usefulness of their results; expressions like "joy," "pleasure," or even "happiness" appear less frequently in this context. It seems that modern scientists no longer experience pleasure in their research and that they have lost their curiosity for the new. The contemporary representatives of scientific knowledge no longer take any pleasure in thinking—an activity that Albert Einstein once dressed up with the word *pleasant*. Yet in the early decades of the twentieth century, it still seemed true that research scientists could look forward to some intellectual fun in getting to the bottom of nature's beautiful tricks. The pleasure of thinking provided the actual motivation for many scientists. Einstein expressed again and again that the form of his research—his formulations and his formulas—got their real beauty when they came about out of the pure pleasure of contemplation, or contemplation driven by its own desire. Max Delbrück as well, who in 1969 was distinguished with the Nobel prize for medicine as a pioneer in molecular biology, could enjoy scientific contemplation in full measure as could many of his contemporaries.

Things are different today. Scientists are no longer the happy individuals they once were. Few of them, in any case, refer in public to the potential pleasure research activity offers, or to the moment of happiness at a pleasing thought. In prevailing numbers they have forgotten what stood at the outset of their pursuit of knowledge, the joy of *aisthesis*, pleasure in the perception of the senses. And they just about notice—in spite of their highly abstract ascetic posture—what they have before their eyes—elegant theories or beautiful structures with potential for aesthetic impact.

It appears that no one in the community of scientists mentions being overcome by emotion anymore when the talk is of scientific insights, and this, I feel, is a striking example of what is missing in science. When we suppress profound emotion, we suppress an intrinsic element of knowledge— today's approach—and cannot realize the experience of beauty integral to a science that raises the claim of being a part of human culture. Without this experience and without this bond, without the aesthetic, sensory, and mood-altering element of science, it is impossible to undertake the new moral and ethical (re-) orientation that seems to be called for, and its necessity will eventually become more pressing.

We have understood this already for several decades—no later than the existence of the atom bomb. The invention of the bomb marks a deep upheaval in the evaluation of science when we look at the history connected with it from an aesthetic viewpoint.

Lack of Reason and Lack of Feeling

About 40 years ago Karl Jaspers wrote his book, *Die Atombombe und die Zukunft des Menschens* (The Atom Bomb and the Future of Man, 1958). It had become clear to Jaspers what had so disturbed him and to what he wanted to call attention, and it concerned the science of nature. He was disturbed by something that lurked in the violent forces unleashed since 1945.

The Allies and their physicists built the bomb at the end of the Second World War for good reasons, and the research scientists delivered technical works as well as theoretical ones that were evidence of the highest ability. The atom bomb had a political basis; it was scientifically and organizationally a masterful achievement and yet "something" was wrong with it, and this is what disturbed the philosopher. He could no longer ward off the impression that the rationality that, up to this point, had been allegedly so superior in steering the development of humanity had suddenly led us in the wrong direction. The ancient prerequisite of any science, that the projected and operative human intellect and rationality produce the good and with every encroachment of nature make living better, was for the first time no longer true. There was overwhelming evidence that the scientific mind had—in the form of the bomb—brought something "evil" into the world. Jaspers suspected that this historical turning point would have an influence far beyond its immediate significance and by necessity would change the scientifically conditioned world.

It was the intellect that had sent humanity down the wrong track. Its authority alone could no longer change anything, and so a means of countering it (and its supporters) was urgently necessary to keep it from producing still more destruction and knowledge of destruction.

Jaspers, who was not only a philosopher but also a doctor, followed a suggestion from therapy. He had a trustworthy talent for guiding the intellect and for giving people the orientation they seemed to lack. This authority was reason, a faculty that Western philosophers have valued since the days of enlightenment. In 1958, Jaspers believed that the future depended on whether humanity would succeed in giving reason a role in intellectual achievement so its accomplishments would once again have a humane meaning and would fulfill a beneficial purpose.

The word *reason* recalls Kant and our awakening out of immaturity in the eighteenth century or takes us even further back to Galileo who, since the seventeenth century, already used reason to exercise criticism of the arrogant authority of higher clergy. When we think of these great names, we feel that we are in good company. The sentimentalism of the Romantics is remote and so Jasper's suggestion can be appreciated.

We attribute many things to reason both then and now and celebrate it as the highest ability of humanity, but otherwise we have gone on with business as usual: We have used our intellect as usual, studied the world as usual, and as usual learned to master it more and more. This is what we wanted and this is what we have been able to achieve.

There was little opposing commentary. By defining reason as simply the ability to exercise criticism and swim against the stream, we saw right away the elementary lack that the resulting development had from that time onward. Reason took place from the very beginning with very little reason, and instead of improving the situation, it grew worse—as people today can somehow sense. People could no longer accept what a few smart people already recognized in the 1960s—the conclusion drawn in the Jasper analysis. In the meantime, the scientific intellect produces more and more "evil" even when it doesn't intend to.

To present only a few examples, in 1963, Rachel Carson realized the imminent threat of a silent spring without birds if we didn't find sensible (and inexpensive) ways to deal with pesticides. Then, Max Born, a Nobel prize winner for physics, warned that sending a man to the moon, though certainly a triumph of intelligence, was at the same time a failure of reason. At the beginning of the 1970s, the Club of Rome recognized that there were "limits to growth" and a short time later an "energy crisis" shook enlightened Western society that in turn became mistrustful and lost all reason.

Suddenly, the parallel arrival of gene technology no longer met with unanimous agreement as the technology of the past had. On the contrary! Its arrival was met with the most severe rejection that remains to this day. Catch phrases like "hole in the ozone layer," "acid rain," and "greenhouse effect" are no longer the only sign that there's to be no more talk of the power of technological intelligence. Technological intelligence must instead admit its impotence, and that it lost all orientation long ago. The "environmental damage" that can no longer be overlooked expresses in words that reason has not achieved what we entrusted it with, namely, the avoidance of destruction that the intellect—the expert intelligence—can give rise to when we do not give its work direction.

Philosophers at the end of the twentieth century have stumbled on this diagnosis without submitting seriously to any therapeutic advice, which has a short expiration date in this media age anyway. The subject often arose that we were learning more than ever how to take possession but were losing all orientation. However, are we humanly capable of recovering it? In the midst of already overtaxed reason, this question is generally passed over with a shrug of the shoulders and then preferably passed on to the manager and the researcher in the industry or the university.

One of the most critical minds of our time expressed a notable opinion on this subject—Wolfgang Pauli, a physicist teaching in Zurich. From Basel, Jaspers sent him a copy of his book. Because Pauli is unfortunately one of the great physicists who has never had a biography written about him, he is known, if at all, only as a physicist among physicists. In fact, Pauli was much more than a mathematically versed theoretician to whom many colleagues attributed even a greater genius than Einstein's. Pauli had a well-rounded education and knew his way around equally well in philosophy, psychology, and philology. He observed and analyzed the world as a critical humanist, as we have been able to read for some years in letters that are available in his multivolume *Wissenschaftlichen Briefwechsels*.

Pauli was the only great physicist of our century who completely kept his distance from the building of the atom bomb and instead solely occupied himself with basic research. He did this not for political reasons but directly from the insight that Jaspers came to later on at the end of the 1950s; when Pauli contemplated the atom bomb, he saw evil in it and worried about the future of humanity.

It is particularly interesting to note what Pauli recommended as the better alternative to reason, which, even after he had heard the same grim diagnosis of scientific rationality, was a completely different prescription for science and for humanity.

As strange as it may sound on the first reading, Pauli believed that the antidote to reason was feeling. He was referring not to that which is understood in everyday speech as feeling, when for example we have an unpleasant feeling in our stomach, when a shirt feels soft against our skin, or when a feeling comes over us, and he thereby also had no irrational form of some kind of easy emotion in mind (as pleasant and exciting as this state may be). Pauli had in mind a well-defined psychic function of humans, that, put more precisely, complements the intellect (thinking) and puts the feeling human being in the position to recognize what is important and meaningful and what is unimportant and meaningless.

The feeling viewed as rational in psychological circles is not concerned with the logic of causes and effects, and it is also not enticed by numbers and data. The feeling with which Pauli was concerned in science recognized the value of something, for example, that of a riverside meadow or a butterfly, and indeed the value that one cannot calculate or grasp with numbers.

The Idea of Complementarity

Feeling cannot be left on its own, and the deciding word is *complementarity*. The idea of complementarity refers to quantities or objects that belong together even though they are mutually exclusive. This idea for

Figure 8.3. The Chinese yin-yang symbol that encloses an s-line in a circle, making it twice as beautiful.

organizing opposites is not new. On the contrary, it runs through the whole of human history. It begins with the Chinese yin-yang symbol (Fig. 8.3) and is not over with the wave–particle duality of modern physics, whereby with this form of complementarity we mean that physicists had to discover that atoms could be waves as well as particles. We also can see complementarity in the idea that in terms of science we can understand nature either as an environment to take advantage of or as "Mother Earth" who takes care of her children.

For any description of nature today, as is my conviction, there is an equal and opposite complement. It is absolutely essential to think in terms of complementarity to understand the whole of the world in which we live where the native goes with the foreign, light goes with shadow, Logos with Eros, and emotion with the intellect, even when, in my opinion, this has escaped notice in the twentieth century.

A look at the history of Western thought shows that after the seventeenth century when the modern form of science gained acceptance, feeling had to withdraw to leave more room for the intellect and its will to power. We researched nature to master it, and in doing so we lost sight of all other goals. The progress of the mind triumphed over the soul and its feeling. At the same time, consciousness succeeded in doing the same. It no longer wanted to be disturbed by unconsciousness. Since then, in a psychological sense, human progress has been more or less blocked. Our development takes place only on the technological level, and the soul is neglected.

However, just because we can't see something anymore doesn't mean it has no more influence; sooner or later opposite forces reemerge. At the turn of the twentieth century, the unconscious emerged again with the help of Sigmund Freud. Even those with biases against certain psycho-analytical schools of thought must admit that the existence of the uncon-scious is not to be doubted, and that by incorporating it, we open up the opportunity for healing an individual's mental suffering.

Thus the unconscious stands opposite the conscious in the same way feeling stands opposite thinking. It can be influential where the tendency to favor the intellect is too constricting and is on the decline. Many people have already made their flight from the region of rationality, be it in some kind of mysticism or mania. Most eventually come back when they notice that both attitudes live within us. Each contains the germ of its opposite, as expressed in the yin-yang symbol of the Chinese tradition (Fig. 8.3).

Reason alone won't help us find a way out of this crisis, but if we have the courage to call on our feeling, we might at least find the road to health. The next time we are told of the profitability of a new violation of nature, we can use feeling to help us explore the complementary opposite. The complement to power and manipulative expansion of influence in gene technology, land cultivation, or the like is what we might call the indif-ferent well-being of a tree or of a landscape. Instead of making a purpose-ful use of nature we can make a purposeless observation of its beauty. Only when the natural sciences value feeling again will they once again value their namesake—what they lost sight of a long time ago—nature. If that happens, science will no longer destroy the environment as a reservoir, but will want to preserve the environment as the foundation of life.

The difficulty of Pauli's advice is that we can't just replace one thing with another and intend to follow it just as blindly, blindly solving every problem of reason through the complementary ability to feel. The difficulty is in keeping both aspects in view at the same time and having to decide each time which is more important, intellect or feeling. The trendy skeptics of science who damn all expert scientific reason lock, stock, and barrel simplify the situation and actually make it worse. By disregarding ration-ality, they rob feeling of its complement and leave it to its own devices. We can't afford to tolerate the resulting escapades any longer. Every form of one-dimensionality chips away at our future. Feeling needs the intellect because it alone knows how and why things are and the intellect needs feeling because without its complement, it forgets what matters and loses courage for the future. Together they could work beautifully.

The Aesthetic Function

From the Bible we know that there is nothing new under the sun. In fact, everything pursued and formulated in this book can be found in a single lecture that Adolf Portmann, the zoologist and cultural philosopher from Basel, gave shortly after the Second World War, in 1949. Portmann spoke at that time of biological aspects in aesthetic education ("Biologisches zur ästhetischen Erziehung"). He reminded his audience that alongside highly praised rational thought and its ability for scientific analysis, which he summarized as the "theoretical function" of humanity, there was also the "aesthetic function" as its complementary counterpart, which he described as having to do with the impressions of the senses, with perception. The aesthetic function is much better at providing the "basis of human behavior" than the theoretical function that, according to Portmann, was responsible for the "crisis of reason" that "Western Civilization" was in and perhaps is still in today.[5]

In the quote at the beginning of this chapter taken from his lecture on the biological aspects in aesthetic education, Portmann tried to awaken the "artistic side" present in all people. In Portmann's texts, which unfortunately are no longer available in bookstores, he says the following:

> Very few have the insight that the aesthetic position needs strengthening—all too many give the pure development of the logical side of thinking the most important task of our human upbringing. This kind of thinking is a sign that we have forgotten that real thought, productive thought even in the most exacting areas of research needs the intuitive, spontaneously creative to work; the aesthetic function, dreams, waking dreams, and all sensory experiences open up inestimable possibilities.[6]

Portmann isn't encouraging science or the humanity of the Western world to replace the exaggerated growth of the intellect with the seeing of visions. He views the goal as "a harmonic balance, a happier person," for bliss or fulfillment in itself and not as a means to any kind of practical end. We can reach the desired harmony by remembering the sensory form of knowledge—the aesthetic—and by giving it a place alongside our logical intelligence, so it at least has the opportunity to exist. The aesthetic function is found in all people, as Alexander Gottlieb Baumgarten maintained in the middle of the eighteenth century when he related his theory of sensory knowledge and tried to give the word *aesthetic* its actual meaning.

Knowledge through Feeling

Around the same time that Portmann was talking about a new way of accessing knowledge in Basel, Pauli was in Zurich trying to bring attention to the dark side of science.[7] Pauli was convinced, based on his own experience, that not only conscious thought can lead to insights into nature, but that dreams and inclusion of the unconsciousness can also help us discover them. Even someone like Pauli, who spent his waking hours at the science institute working with complicated mathematical calculations immersed in all things rational, began to see that the most simple connection to the unconscious is the human psychic trait known as feeling. This feeling, in the scientific context, has nothing to do with the everyday understanding of the word such as feeling well or feeling an itch or whether a shirt feels soft. (This does not mean that we use the words *to feel* or *feeling* in the wrong way in the everyday context, but that listening to the context is essential to avoid misunderstandings. On the other hand, it would be nice if the *feeling* I am referring to in scientific terms were as popular as everyday feelings.)

The feeling that is critical to the sciences is the ability to recognize what's important (which almost always keeps us from error). This sense is necessary to access Portmann's aesthetic function. Heisenberg, for example, had this feeling as a young man in his ability to see the beauty in symbols of physics deep within himself. Einstein talked with psychologists about a "feeling for direction" that gave him the strength to wrestle with all of the mathematical problems that were still in transition, from their initial images to articulate insights ready for the public.

Pauli reported a strange feeling he had in the 1920s that allowed him to understand the properties of an atom that give it a further degree of freedom that today we call *spin*. At that time we knew that electrons, as all particles, had three degrees of freedom of movement that allowed them the three dimensions of space. They could change their position forward or backward, right or left, and up or down. Quantum theory of the atom established these possibilities through quantum numbers, and at the time there were three. That was the end of it until Pauli realized there had to be another property of electrons involving turning on their own axes, which was a different movement from the others. He gave them a fourth quantum number and formulated his "exclusion principle" that keeps electrons in an atom or molecule from simultaneously occupying the same quantum state.

Although it wasn't hard to convince his colleagues with his mathematical proof that operated to a great extent according to the concept of symmetry, Pauli didn't let on "what his real work was": "wrestling with his naive 'intuition' that the fourth quantum number had the property of one and the same electron" as he once wrote in a letter to his colleague Markus Fiery.[8]

Here is an example of the "spontaneous creative work" of which Portmann spoke: Pauli took the spontaneous step from three spatial movements to four quantum numbers unconsciously without using logic, just as Heisenberg unconsciously and without intention caught a glance of beauty. Heisenberg appears to simply have had a revelation, but Pauli can be classified in a larger aesthetic framework, based on his own indication of the "difficult passage from three to four" that he had to make and the passage at the beginning of the seventeenth century that went in the opposite direction.

Trinity and Quaternity

It doesn't take long to run into difficulties with figures and the meanings given them because the scientific tradition that has ruled Western thought since Bacon, Kepler, Galileo, and Descartes consists of measurable quantities, and because the meaning of the figures is limited to this form of information. Numerical figures no longer express any kind of quality in today's usage, at least not in the scientific sphere. Numbers stand for the results of measurements, and these results are our main concern.

Kepler saw things differently. For Kepler, three was not just any number. It was a holy number that expressed the three-in-one (trinity), and he came to scientific insights and convictions regarding it. Because four never came up in his thinking, in spite of his precise knowledge of the Pythagorean worship of *tetraktis*, and we know that this "valued four" was in high standing in all his procedures, we can conclude that something fundamentally new was happening with Kepler. Apparently he had a new "spiritual attitude" that "with significance far beyond Kepler's person, produced the natural science that today we call classical."[9]

This is how Pauli formulated it in an investigation from the early 1950s that is far too little known and discussed in science on the influence of archetypal images in Kepler's theories, *Einfluß archetypischer Vorstellungen auf die Bildung wissenschaftlicher Theorien bei Kepler*. This spiritual attitude brought about the end of the old alchemical thought that held that in all matter there sleeps a world-soul, the *anima mundi*, responsible among other things for the generation of new stars.

"Three" created the archetypal basis for the successful development of the natural sciences in the century that followed. This was Pauli's conviction and in this context it becomes clear what the passage from three to four signified to Pauli. Quantum theory's requirement for a fourth parameter for an electron signified a transition to a new archetype, one that would determine modern physics.

No Sense

The fourth component is what science, though hardly ailing, lacks, and it presents us with the challenge to take the previously so successful application-oriented procedures and arrange them to protect its known qualities and enrich them with new ones.

Traditional physics is trinitarian with its three basic dimensions: the space–time continuum so clearly indicated by Einstein, indestructible energy expressed in the nineteenth century, and causality that determines all activity in the space–time and energetic environment. What is apparently missing is a counterbalance to causality, its complement, and this lack must have something to do with what we call sense, sensuality, and sensory perception.

Traditionally science attempts—as in its attempt to explain biological evolution—to place causality in opposition to chance, or to the side, and we are quick to consider the pair of concepts, chance and necessity, as ultimate wisdom. Yet this approach seems to me to put too much of a burden on the real, existing chance in several ways. For one, chance, as in variations that appear by chance, seems to play an almost exclusive role in evolution and one that fits in with the whole of nature, which means it makes sense. This is hardly pure chance. And for another, feelings make many people aware of their dissatisfaction with this interpretation of their origins. Isn't there something else between the loving God and the off-chance?

What the whole traditionally trinitarian natural sciences are lacking under the leadership of physics is an understanding for individual phenomena. Baumgarten referred to this point in his *Aesthetica* of 1750, when he discovered that abstraction—in all the brilliance of its theoretical ability—is a loss, a loss of the individual. Trinitarian science knows only the general: physics, the general instance of a motion; botany, the general knowledge of a plant; medicine, the general knowledge of a patient. The interchangeable is handled within this measurable framework in accordance with natural principles. As a result, its complementary elements disappear, that is, the unmistakable—and what is provided by chance. Neither one even comes into view. Rationality has no place for it, and therefore it must be supplied (complementarily) by the sensory, by perception, this being all we have available for taking in the particular and the particulars.

It is precisely this inability of modern physics to grasp the specific instance—the motion of a single electron, for example, or the decay of a single atom—that Einstein called to account with his criticism of the mechanics of the atom. He believed with unshakable conviction that quantum theory was not complete, and defended this to his dying day. He dreamed of a solution in the classical form of his science, and set his heart on the old forces of natural law, a need that Pauli once wrote sarcastically about, calling it "Einstein's neurotic misunderstanding."[10]

Physics can probably only achieve completeness in the sense of complementarity—along with all exact sciences—when it includes sensory cognition in its consideration of reality, making use of the atrophied aesthetic function, as Portmann would say. I presume that in this complete (perfect) knowledge, beauty is showing human beings what really matters. Our undeniable "artistic impulse" demands it. We want not only the One that is All (rational abstraction) but also the One that is just one and can be known individually.

Aesthetic and Moral

Beauty has everything to do with perfection. At least Baumgarten sees it this way, who, again, was the first to attempt opening the scientific eye to beauty. In a 1996 lecture on the aesthetic foundations of morality, *Über die ästhetischen Grundlagen der Moral*, the young philosopher Michael Hauskeller indicated that going along with Baumgarten's program would have decisive moral consequences, for "Beauty makes reality something of value: It attracts our undivided attention and makes us bashful. The gap between what is and what should be that . . . generally seems inconquerable may very well be closed by aesthetics."[11]

The gap between what is and what should be, this dogmatic syntax of the old school of a value-free science that sees no way to bridge the two concerns, does not exist for the perceiving person. All parents know when they look at their child that it is up to them to help and to be responsible for that child, and not only parents sense this. "In the phenomenon [of a newborn child] what is and what should be are inseparable," [12] wrote Hauskeller, who learned this from Hans Jonas. His "imperative of responsibility" finds "the elemental 'should' in the 'is' of the newborn," whose very breathing demands from its fellow human beings that it be taken in as their own.[13] Jonas summarized this insight with the fine words, "Just look and you know." Hauskeller used the example of the Milgram experiments, one of the classics in psychology, to demonstrate the extent that perception, the fulfillment of the aesthetic function, is what enables people to behave morally.[14] The Milgram experiments were carried out at the beginning of the 1960s and since then their results have been confirmed many times.

The American psychologist Stanley Milgram wanted to know how adults responded to obedience to authority and set up the following situation: He employed two actors in his laboratory who played a teacher (as an authority figure) and a student. They were ostensibly trying to establish whether punishment heightened success in learning.

Enter the real test subjects—the infamous men and woman on the street. They were asked to provide assistance in a pedagogical experiment and regarded the teacher and student as authentic. Their task was simple: All they had to do was carry out punishments prescribed by the teacher that were designed to improve learning. The punishments were incremental electric shocks that could be discharged at the push of a button (with dummy equipment). The test subjects could deliver up to 400 volts, a deadly dose if they were really to carry it out.

Milgram's question was whether people under the observed influence of authority (of a professional teacher) would carry out punishment they themselves must recognize as insane and disproportionately brutal. The distressing answer given by the resulting picture of the middle class was unfortunately a persistent "yes." Because this aspect is already extensively covered in other literature, I would rather discuss an observation that came out of the Milgram experiments and that fits into the context of aesthetics. This observation demonstrated that the readiness (the obedience) to carry out extreme and life-endangering punishment depended on the proximity of the test subject to the punished "student" (Fig. 8.4).

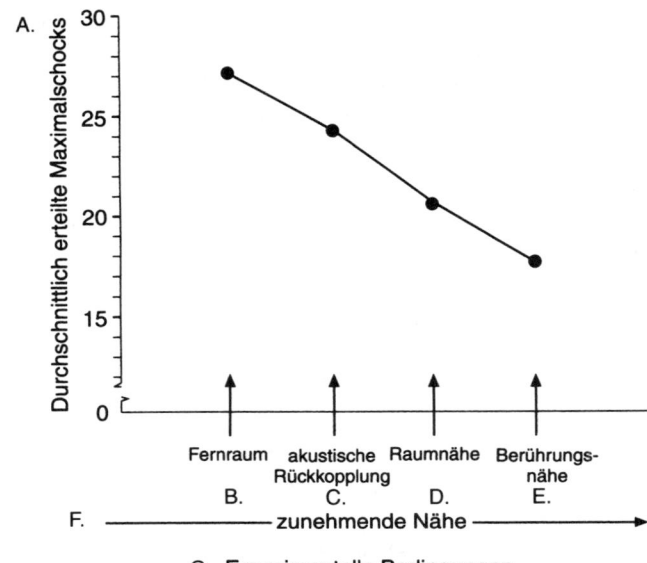

Figure 8.4.   The Milgram experiment: Whether the maximal punishment was meted out depended on the spatial proximity of both persons. The individual positions are described in the text. A. Average maximal shock given; B. Remote; C. Voice feedback; D. Proximity; E. Touch proximity; F. Increasing proximity; G. Conditions of the experiment.

Milgram distinguished between situations where the test subjects and the "students" were in separate rooms and situations where they conversed in the same room. In the first case, he distinguished further as to whether or not the test subjects perceived nothing about the "punished" ("Remote") or whether they heard them (groaning or crying)—("Voice"). In the final experiments, he once again distinguished whether or not they could see each other ("Proximity") or whether or not they even had the opportunity to touch ("Touch proximity").

    The results showed that the voice feedback changed almost nothing in the behavior of the test subjects. The number of obedient subjects fell only from 65 to 62. Only "Proximity" had a significant influence—the number of obedient now decreased to 40—and sank to 30 with the increase in bodily contact. In the end, the overall high degree of readiness to unqualified obedience is disturbing, but only one aspect of this finding relates to our current subject. The Milgram experiment took place in a cool and sober room (a laboratory), a room devoid of aesthetic charm. I think the question of how the test subject might respond in an aesthetically pleasing environment, which would include artistic creations accessible to the senses, would be worth a Milgram experiment, part II. In this second part, Milgram would test whether the subjects would still press the dangerous buttons if they were able to see Winslow Homer's *On a Lee Shore*, for instance, or *Starry Night* by Vincent van Gogh.

The experiment took place without art, however, and only these results are known (Fig. 8.4). We might have expected the general tendency toward obedience, but we can still gain something from thinking about it a little further. Its significance for aesthetics is that as the punished person was situated nearer and nearer to the test subjects, the subjects weren't learning anything conceptually that they didn't know before. They were already aware that he was in pain, that he was suffering, and that he was a helpless creature well before he came into their field of vision. They knew it, and at the same time they didn't know it. The test subjects—people like you and I—at first only knew through concepts what they were later to perceive and understand with their senses, and only when they saw, did they know—exactly as Hans Jonas put it. They knew at once that it was a human being they were causing to suffer, and they knew that it was this particular human being sitting in front of them they were causing to suffer.

Aristotle characterized the knowledge supplied by perception as knowledge of the particular, and it is only this knowledge that grasps reality. The general is unreal and leaves our morality cold. Science that operates exclusively according to the theoretical function does not reveal anything that would lead us from what is to what should be. It's a little different with the particular, and when I manage to perceive its beauty, I have a pretty good chance of knowing what I should do. The "is" perceived by sensitive human beings leads to "should."

A Contemporary Field Trip: The Clone of No Value

It is perception, then, that is the starting point for the development of ethics in science. Sensory experience is able to convey the sense of value that leads to a moral attitude, and it does this because, according to Hauskeller, it lets "the things have their souls."

The concept of the soul in modern science is apparently dead on arrival, and quoted again and again are the words of Rudolf Virchow, the famous pathologist of the nineteenth century, "I have opened up thousands of corpses and never found a soul." However, the soul maintains its influence apart from this quip, and even if many scientists deny its existence, most people do know that they have a soul. It might not be such a bad idea to imagine the soul as what people see when they perceive another human being. When we observe someone else, we focus our attention on certain parts (hair color, height of the forehead); when we perceive someone else, however, we grasp something beyond this. We could (and I would) call this the soul.

The soul is related to a contemporary subject that fits into the broader context of aesthetics—the issue of cloning living creatures. By cloning, I refer to the current capability of scientists to manufacture identical copies of genetically advanced higher forms of life through asexual reproduction. In March 1997, "Dolly" the sheep was produced by cloning; primates are apparently next and the public is waiting in suspense for the first human clone.[10] There is already a business founded in the Bahamas for the purpose of cloning human beings that has found an eager clientele among people who apparently consider themselves irreplaceable. The idea of watching themselves growing up in the twenty-first century must fascinate them.

There is a lot of excitement surrounding this subject, and many voices have surfaced that argue for outlawing human cloning on the basis that it is reprehensible. Somehow this all seems like old hat. The Christian church, for example, is talking about the unlawful violation of nature once again without considering that people have never done anything but this. In addition, "the creation" has turned into a mere concept for theologians, and it has been a long time since they've seen it as a way of viewing life, never mind as having anything to do with perception.

Many fear a human clone because they can no longer overlook the fact that Dolly represents a blatant lack of ethical criteria. How can we find our way? The old wisdom of the scientist who operates rationally and produces only good, his scientific intellect never running into ethical problems, is gone forever. In addition, no one seems to have come up with a mechanism that can or should take its place.

Everything science has done for hundreds of years has flown under the banner of the grand "idea of progress," which proclaims that what can be planned using scientific intelligence is for the good. It is in this context, then, that someone can ask why we shouldn't try to clone people one day. We don't have to look far to find cases where utilitarian imagination makes this possibility conceivable. In fact, the reasons are at hand—in the narrow context of the childless industrial worker who has found the way to hand his business down to (freshly cloned) hands, or in the broader context of governments that have an endless need for young who are strong in battle and weak in thought to fill the lower ranks of the military. Why shouldn't—someone might ask—anyone with enough money try to clone the needed entrepreneur or the desired underling? Someone somewhere will build that person a laboratory, and if yesterday speculation was drawn toward the Brazilian rain forest, today it's the Iraqi desert. Why not, when the risks are so minimal? What can go wrong? If we want to clone people, shouldn't we at least try? Is there a problem with that?

Well, there is the risk that some of the desired traits, such as business intelligence or the soldierly willingness to fight, have nothing to do with genes but are determined by the environment. In that case, cloning would fail for lack of a contingent environment. Even if the parents and the grandparents of the manufacturer were cloned to provide the necessary environment, they too would only be identical clones of their ancestors, even if their parents or grandparents had also been cloned along with them. It doesn't take long before we are lost in a fog of impracticability.

Then there is the risk that the aging businessman who has been reproduced would be completely unsuited to do what needs to be done in the new era. He may have been able to do a great job of founding a typewriter or a cash register factory in his hometown and expanding nationally, but it doesn't mean that he has grown up adapting his work to the changing demands of the open market.

Similarly, as far as the soldiers are concerned, they're bound to arrive too late (if they are going to be manufactured with genetic material and not with brainwashing propaganda as the Nazis did). Cloning would require decades, not months, and who can guarantee that the political situation in the meantime—the image of the enemy—would remain stable for that long? We only have to imagine one of the great powers in 1975—still in the fading cold war with millions of contracted soldiers—wondering what to do today with all the superfluous human clones after the fall of the Berlin Wall and the dissolution of the Eastern bloc. At the very least, the governments would have to create jobs for them.

There is also another risk: that the passage from sexual to asexual reproduction would be detrimental to the genetic variety of the human

178 Beauty and the Beast

species. Cloning could diminish the human ability to react genetically to sudden changes in the environment.

Although the subject is important as a matter of principle, apart from any technical considerations, it is extremely unlikely that people will be cloned to any great degree. However, this improbability has nothing to do with our fear of the possible biological dangers. Pure and simple, there is nothing tempting about the prospect of an earth populated with genetically identical people, as enticing as it may at first seem.

At first glance, it might be amusing to see Immanuel Kant, Johann Heinrich Pestalozzi, Ludwig van Beethoven, Albert Einstein, Marilyn Monroe, Empress Maria Theresa, or whoever it may be suddenly appear and stroll around on our streets in large numbers. However, if we take another look, we'd have to imagine that these creatures could no longer do that for which we hold them in such high esteem. Einstein would no longer discover his theories or Beethoven his symphonies, and the assumption that Beethoven would produce completely new music and Einstein would pursue a new science is more than questionable. It is more probable that both clones would experience their existence as useless. Readers can fill in the details of the psychological consequences.

If we go to the other extreme and imagine that Einstein would master all forms of modern science and Beethoven all techniques of new music—with the genius we expect from them—then all those in possession of a normal intellect would be left with only resignation, to put it mildly, or with a bullet.

This leads us to Marilyn Monroe, who, as we all know, committed suicide. Everything we know from her biography indicates that although she looked happy enough in her life, she was not happy. Why is it then that when we talk about cloning, we talk about cloning her?

I doubt that Ms. Monroe would agree to live a reconstituted life, and I doubt very much that she would do this before a grotesquely curious public who wouldn't consider for a second giving her a chance to live in peace. In my opinion, Ms. Monroe would much rather stay dead than be further victimized by the day's gossip columns.

It doesn't take us long to lose all desire for cloning once we take most of the individual instances into consideration. However, there is something even more important that has to do with the idea of seeing the very much desired human beings, but seeing them in the hundreds or thousands, that is, copies. The image quickly loses its intoxicating effect and instead leaves the worst possible impression. As humans, we are frightened by a crowd of completely similar looking serial productions. Why this is true has to do with what has been called the uniqueness of the human soul.

When we cannot see evidence of uniqueness in a human being, we do not see that person as a human being, but more like a cold Cartesian machine with which we can do anything we please. Anyone can make a study of depersonalization by taking a good look at racial hatred or xenophobia or by asking why, when people are wearing soldier's uniforms and are not recognizable as individuals, they become easier to kill. Every faceless mass—the stranger or the soldier—is soulless and can be declared anything from a collective enemy to a beast, with the basic tenets of morality having nothing to say about it. In other words, the cloned human being would not be perceived as human, and if we encountered such a creature in duplicate, we would probably end up killing it along with its companions. I for one would be terrified by my own clone.

We are now honing in on the crux of the debate and the question that asks what the actual source of moral notions for the Western scientific world should be. What is there to stop us, really, from letting even more "evil" loose in the world—from cloning human beings, for example? It's difficult to find a solution in the scope of present-day science that deals exclusively with the measurable and has no approach for unique human beings who might be capable of losing their souls in the cloning process. I have tried to indicate in this book how we can broaden the scope of science to find a way out. When rationality runs up against its limits and recourse to enlightened reason is no help anymore, then it may help if we become conscious of our given human ability that serves as a counterpoint to thinking—feeling. Feelings are released through our senses; they help us to perceive the world and recognize value.

I am talking about value that cannot be expressed through numbers—a price to be paid, for example. The human clone represents this kind of quantitative value, and it is this quantitative value that is responsible for cloning in the first place. I am referring to a value that can't be measured or calculated. I am referring to a value that we can only comprehend through sensory experience or perception, and one example of this is what we understand as beauty.

The concept of beauty has been applied so arbitrarily, in fashion journalism or by the cosmetic industries, for example, that it has become devalued. It still belongs at the center of science, however, where scientists need to remember that although they are always trying to expand their knowledge of nature to gain more power over it, they are also studying nature out of a sense that it is beautiful. Whatever gives a thing its perceptible value, we call beautiful apart from any application.

The "atrophy of the life of the senses" in the sciences must come to an end, as Portmann emphasized in the 1950s. The aim of science must go beyond mere usefulness when the actual source of its existence lies in the power to perceive. Science must find its way back to this source, and if it succeeds and in this respect becomes something of value again, then no one will be interested in cloned human beings anymore. They will be of no value.

Epilogue
Ending with a Dream: The Scientist and Beauty

All the world understands the language of beauty.[1]
Peter Sitte

In the *Notebooks of Raymond Chandler*, inventor of the legendary Phillip Marlowe who knows there is no justice to be found on the streets of Los Angeles, there is an entry for February 19, 1938, that's completely different from the other entries and comes under the strangely conspicuous heading, "Great Thought":

> There are two kinds of truth: the truth that lights the way and the truth that warms the heart. The first of these is science, and the second is art. Neither is independent of the other or more important than the other. Without art, science would be as useless as a pair of high forceps in the hands of a plumber. Without science, art would become a crude mess of folklore and emotional quackery. The truth of art keeps science from becoming inhuman, and the truth of science keeps art from becoming ridiculous.[2]

The entry seems disconnected from the others. Chandler must have been gripped suddenly by the "great thought." Maybe it came to him as an inspiration, an inspiration for the future. My dream is that both human efforts can be united around the truth in the endeavor to find beauty. Couldn't the dream be realized if the old science and the old aesthetics met and founded a new, aesthetically oriented science? Aesthetics without science is useless and science without aesthetics has no value. Science with aesthetics can be of value. What is a part of science and has been missing up to now is the viewpoint of art in the sense that art serves as an

181

interpreter of nature. What the scientist is lacking is the perception of the artist who studies nature to present experience in a new form.

"Ironic Science"

The dream of a new beginning for science comes at a time when many are complaining about the end of science. *The End of Science* is also the name of a book published in 1995 by John Horgan.[3] In it, Horgan interviews prominent end of the twentieth century scientists, such as Roger Penrose and Freeman Dyson, and asks them for their judgment concerning the future possibilities or limitations of science. One issue that surfaces is a strange difficulty reported by both theoretical physicists and their mathematically versed colleagues in cosmology. These scientists are experiencing the strange phenomenon that they are able to provide theories on things and the world as always, but they no longer have any opportunity to prove their theories through experimentation. The equipment needed and the energy required exhaust all earthly possibility. In other words, their current theories have about as much empirical evidence as the statements of science fiction authors or of other writers. The same statement can be formulated with an unusual perspective: The difference between science and literature is gradually disappearing. Put positively, the public could judge what a theoretician of the universe says about the world according to the same criteria used for theories conceived by a writer. Everything would then depend on form, and would enable science to finally become aesthetic.

At the moment, science is not aesthetic. There are some scientists who do recognize the dilemma as outlined here, and one of them, Christopher Langton, the chaos theory scientist, once suggested that there is "something more like poetry in the future of science." However, the conclusions that some of today's stars draw from the facts, that their statements are as provable in experiments as the lines are in poems, sound a little sullen. Stephen Hawking and others, who like him are searching for black holes, no longer take themselves seriously and call what they are doing "ironic physics." No one really understands their technical mumbo jumbo (Fig. 9.1a), but even the lay reader can easily recognize that on the aesthetic niveau it's a disaster (Fig. 9.1b).

Ironic science means the end of science. The only way for scientists to find their way out of the dead end in which many branches of science are

$$
a. \quad \left\{ \begin{array}{c} \phi_{A'\cdots G'}(r) \\ \text{or} \\ \phi_{A\cdots G}(r) \end{array} \right\} = \int_{\omega = i r \pi} \left\{ \begin{array}{c} \pi_{A'} \cdots \pi_{G'} \\ \text{or} \\ \frac{\partial}{\partial \omega^A} \cdots \frac{\partial}{\partial \omega^C} \end{array} \right\} f(Z^\alpha) \pi_{E'} d\pi^{E'},
$$

b.

No-hair theorem. Stationary black holes are characterized by mass M, angular momentum J, and electric charge Q.

Figure 9.1   (a,b) The discrepancy between the technical know-how and the aesthetic failure of the modern physicist is demonstrated in both illustrations from the book *The Nature of Space and Time*, coauthored by Stephen Hawking and Roger Penrose. The reader finds incomprehensible integrals (a) and fatuous jokes (b) juxtaposed. The drawing is supposed to illustrate the so-called "no-hair theorem" according to which stationary black holes are characterized by their mass, their angular momentum, and their electric charge. It couldn't be less helpful to the experimental physicist. Princeton University Press, Princeton, 1996.

stuck is, in my opinion, by embracing an aesthetic science. We can find our way there, if we keep in mind that the essential theories and models of science first made an impression on the observer because of their beauty.

"Beauty Will Save the World"

It will have to come down to an interplay between art and science, if only, as Raymond Chandler feared, because the truth of science has long since turned inhuman. Ethical concerns about morally dubious research have multiplied to the point where even higher powers have been called on to save us from them. Martin Heidegger once said that only a god could save us, and Hans Jonas said something similar, to great public acclaim. In his book on the imperative of responsibility (*Das Prinzip Verantwortung*), he suggested that there must again be "awe for what humans were and are." Awe and dread need to be learned again so they can protect us from the delusions of power that the natural sciences make available. To Jonas, awe will reveal something "holy" that will not become "hurtful under any circumstances."[4]

To me, this seems to be reaching a little too high; instead of straining toward heaven, I would rather keep both feet firmly on the ground. Instead of awe for the holy, I would rather consider the feeling of being deeply moved in the face of nature's beauty that Nicolai Hartmann defined in his *Ästhetik*. Hartmann remarks that the perceptive human being "cannot guard against the feeling of suddenly being face to face with the miracle of creation."[5] For Hartmann, it's clear that the scientific spectacle can be aesthetically fascinating when natural scientists realize the depths of beauty.

This wakening of consciousness could finally lead humans to become moral, according not to Hartmann but to Joseph Brodsky, who died in January 1996. In an only recently published essay from which I have already quoted, Brodsky talks about how each new aesthetic reality defines the ethical, because "aesthetics is the mother of ethics." He distinguishes the concepts "beautiful" and "not beautiful" as aesthetic concepts that precede the categories of "good" and "evil." According to Brodsky, everything is not permitted in ethics because everything is not permitted in aesthetics, just as the color scale of the spectrum is limited.[6]

At the outset of the perceptive and sensual life, Brodsky wrote in his essay, there is an aesthetic choice, and men and women choose the beauty that they understand. Moral ideas flow out of this perception when we turn it toward other people, and it is this perception of beauty, the sensory recognition of reality, that deserves some thought. Brodsky believed that the richer an individual's aesthetic experience was, the more unwavering his taste, the more precise his moral judgment, and ultimately the greater his independence would be. He called for understanding Dostoevsky's utterance that "beauty will save the world" or Matthew Arnold's idea that we will be saved by poetry in both a practical and platonic sense because although it's too late to save the world, the individual can always be saved.

Aesthetic sensibility expresses itself spontaneously; even if we don't know who we are or what we need in reality, we always know what we like and what rubs us the wrong way. Anthropologically speaking, Brodsky pointed out that humans were first aesthetic beings before they were ethical beings and emphasized that art and especially literature are not by-products of the development of our species, but the other way around, "for what distinguishes us from the rest of the animal kingdom is precisely the gift of speech" and in this way literature, and especially poetry, "is the supreme form of human locution in any culture," put simply, the distinction of our species.[7]

The Two Cultures

I find what Brodsky has said here poignant and fitting, and it leads me to point out a troubling problem in our culture too rarely investigated in any depth. I am referring to the rift between two cultures of natural science and the humanities and between the scientific and literary intelligentsia to which British physicist and writer Charles Percy Snow referred in a lecture in 1959, entitled "The Two Cultures and the Scientific Revolution." This division is unfortunately no "Snow from yesteryear," as it states in an unfortunate pun, but instead is practiced in real life. In response to the poet Brodsky's similar point, and to his assertion that literature and in particular poetry is "the supreme form of human locution" and "what distinguishes us from the rest of the animal kingdom," the natural scientist would have to protest.[6] He might ask why the mathematical language of theoretical physics or the symbolic language of analytical chemistry or even the jargon of modern science doesn't deserve this recognition. These too are "supreme" forms of "human locution" that could be understood as "what distinguishes us."

Of course scientific and literary language usage differ, but the difference is not as easy to pinpoint as is often thought. The difference does not mean that everyone understands the language of the poet and no one understands the language of the scientist. I doubt very much that more people can understand a poem by Paul Celan than one of Einstein's formulas. The difference has more to do with what we are astonished by in the respective fields. We are more astonished by the form of a literary text (the work) and by the content of a scientific text. In literature, what's important is the how and in science, it is the what. In physics, chemistry, or biology the concepts are important, the specific terminology that houses the knowledge, for example, of the gene, the atom, energy, metabolism, a wave, or equilibrium. It has to do with knowing what is meant. In literature what is written about is less important than how it takes place and how it turns out.

Science can be viewed as storing its knowledge in formless concepts. It's easy to see through the error in this thinking when we become aware that although scientists are always chasing after the what—What is life? What is matter? What is energy? What is a gene?—they'll never get to the end of the answers. One of the biases against the natural sciences is that they deal with answerable (read: banal) questions. Natural scientists can discover what something is (and they only get better and better at it) but they haven't been able to say what that something is for a long time. In the words of Niels Bohr on the quantum and the wondrous physics of the atom, "Quantum theory is a wonderful example of the way we can talk in complete clarity and understanding about the facts and all along know that we can only talk about them in images and metaphors."

We can also do this in poetic form, as Richard Feynman attempted once when he tried to see what actually happens when gravity unfurls its influence and makes two masses mutually attract each other. The physics presents no theoretical problems. However, how do we express this in regular language without losing what makes the subject fascinating and the problem interesting? Feynman tried his hand at it with a haiku-like verse, as follows:

> Principles
> You can't say A is made of B
> or vice versa.
> All mass is interaction.

Even natural scientists sometimes have to (and are sometimes able to) display their findings in a form more familiar to us from literature. However, when linguistic and other efforts at communication in the exact sciences and technology become more aesthetic, then these fields should begin to keep better step morally. Now as ever, the natural sciences are a technical culture and because they take on responsibility, they respond with one answer. They can, however, become an aesthetic culture that responds with a variety of answers, and become more and more like belles-lettres. In the meantime some physicists, biologists, and chemists do formulate the truth without losing its mystery. The truth is always more appealing in the right form. After all, the mysterious is the most beautiful part of life, as Einstein once expressed it.

The New "Universitas"

I am not implying that science and the humanities should become a single culture. I mean only that the natural sciences should become what the English words *moral science* express, related to the word *humanities*. Science and the humanities will always be different from each other. In one culture we talk about things that can be agreed on—the temperature of water or the energy of a feather—whereas in the complementary culture the first order of business involves the things that mean something to people—joy, affection, and love, for example. If there is an urgent job for cultural institutions today, it is to come to an agreement that both the sciences and the humanities mean something and hold value for humanity.

Is there any hope that our present system of science will allow this? The answer in my opinion is "no." University departments are too fragmented and don't offer the opportunity for finding a "harmonic equilibrium of the intellectual functions" of which Adolf Portmann spoke. We could also say that today's universities no longer offer *universitas*, suggested by its name. *Universitas* refers to the unity of scientific disciplines as well as the unity of people occupied in them.

This dream might be fulfilled if there were a place where higher level science and art were taught together and one would never appear without the other. It should be a place where the students and teachers would be able to seek and find their own personal approach to perception by working in both areas of human endeavor. This is the only way we can replace the former expertise with epistemological methods that allow us to understand the value of nature and its beauty without abandoning the singularly solid ground of scientific method. Aesthetics with an emphasis on perception is this method.

A Personal Addendum

The beauty of this dream is that such a place does exist. It's called Holzen and is in Markgräflerland in Germany. Here a private study college has been founded in the hopes of becoming a school of perception. I quote from the manifesto of the school that up to now has never been published: "Holzen as a university of scientific research would investigate the whole range of aesthetics in order first to approach more closely reconciliation and collaboration of disciplines, and second to overcome the division of experience and knowledge which has led to the malaise of modern science."[8]

Anyone who steps into the recently finished building of the school can read on the foundation stone Immanuel Kant's words that express the common ground for which Holzen strives: "The human being is made to live in society with other human beings to cultivate itself within it through the arts and sciences."

Endnotes

Preface

1. The Goethe (1805) quote is found in his *Winckelmann-schrift*.
2. Heidegger's unfriendly remark was quoted in R. Safranskis's (1997) biography *Ein Meister aus Deutschland* (p. 179). Frankfurt: Fischer TB.
3. Ernst Haeckel's (1974) image is the edition taken from *Kunstformen der Natur*. New York: Dover.
4. The fairy tale or myth is told according to the version in C. G. Jung's (1983) book *Der Mensch und seine Symbole*. Solothurn.
5. The closing quote is found in Josef Popper's (1901) work, *Die technischen Fortschritte und ihre Ästhetische und kulturelle Bedeutung* (pp. 4–5). Dresden: Verlag Carl Reiser.

Chapter 1

1. The quote from St. Thomas Aquinas is found in *Summa theologica* I/II, Question 27, Art.1.
2. Copernicus, N. *On the revolutions of the celestial spheres*, as quoted in Kuhn, T. (1957). *The Copernican revolution* (p. 179). Cambridge: Harvard University Press.
3. Ibid., p. 180.
4. More on Johannes Kepler and Richard Feynman can be found by the interested reader in Fischer, E. P. (1995). *Aristoteles, Einstein & Co.* Munich: Piper Verlag. The quotes from the works by Kepler are found in the long essay by Wolfgang Pauli, "Der Einfluß archetypischer Vorstellungen auf die Bildung naturwissenschaftlicher Theorien bei Kepler" in the book coauthored by Jung, C. G. (1952) *Naturerklärung und Psyche*. Zurich: Rascher Verlag, which contains precise references to *Mysterium cosmographicum, ad vitellionem paralipomena* and *Harmonices mundi*.
5. Ibid.
6. Ibid.
7. Ibid.
8. Gleick, J. (1992). *Genius: The life and science of Richard Feynmann*. New York: Pantheon Books.
9. Fischer, E. P. (1995). *Aristoteles, Einstein & Co.* Munich: Piper Verlag.

10. Fischer, E. P. (1996). *Einstein—Ein Genie und sein überfordertes Publikum*. Heidelberg: Springer-Verlag.
11. The quote from *Faust* is lines 434ff.; L. Boltzmann's (1979) speech is found in *Populären Schriften*. Braunschweig.
12. Henri Poincaré's (1996) remarks originate from his book *Science and method* (p. 22). Bristol: Thoemmes Press.
13. Ibid.
14. Aristotle (1928). *Metaphysics*. Tr. W. D. Ross. Oxford: Oxford University Press.
15. Ibid.
16. Boltzmann, L. (1979). *Populären Schriften*. Braunschweig.
17. The work by Watson and Crick (1953) was published in the journal *Nature, 171*, 737–738. There is also a second work by the duo in the same volume on pages 956–966.
18. Paul Dirac's remarks and others on Einstein are found in my book, Fischer, E. P. (1996). *Einstein—Ein Genie und sein überfordertes Publikum*. Heidelberg: Springer-Verlag.
19. The classic work by Meselson and Stahl (1958) was published with the title, "The replication of DNA in *Escherichia coli*" in the journal *Proceedings of the National Academy of Sciences, USA, 44*, 671–682.
20. The analysis of Robert Millikan is found in the essay, "On the art of scientific imagination" by Gerald Holton (1996, spring); it appeared in the journal *Daedalus*, 183–208.

Chapter 2

1. The quote from Dürer ("Schönheit—was das sey, weiß ich nit,/obwohl sie vielen Dingen anhanget.") taken from the book *Schule des Sehens* by Hildesheimer, W. (1996). Frankfurt: Insel.
2. The classic texts discussed (Plotinus, Burke, Kant, for example) are found in excerpts (and provided with commentary) in the volume, Hauskeller, M. (Ed., 1994) *Was das Schöne sei*.
3. The book *Schönheit—Ein Versuch* (Haecker, T.) published in 1936 is referred to by Löw, R. (1994). *Über das Schöne*. Stuttgart: Weitbrecht.
4. The classic texts discussed (Plotinus, Burke, Kant) are found in excerpts (and provided with commentary) in the volume, Hauskeller, M. (Ed., 1994) *Was das Schöne sei*.
5. The quote from the Taoist philosopher Zhuangzi is taken from Schleichert, H. (1997) *Wie man mit Fundamentalisten diskutiert, ohne den Verstand zu verlieren* (p. 30). Munich: C. H. Beck.
6. The classic texts discussed (Plotinus, Burke, Kant) are found in excerpts (and provided with commentary) in the volume, Hauskeller, M. (Ed., 1994) *Was das Schöne sei*.
7. Darwin, E. (1974). *Zoonomia: or the laws of organic life*, Vol. I/II, London: J. Johnson, p. 568 and p. 722. Reprint 1974, New York: AMS Press, pp. 145–146.
8. A. G. Baumgarten's aesthetic was discovered mainly by Hans Rudolf Schweizer (1967) in *Vom ursprünglichen Sinn der Ästhetik*. Zug. Baumgarten's text itself is out of print, but can be found in libraries. H. R. Schweizer edited some of his works in Meiner Verlag.

9. Hans Blumenberg, as quoted in Jäger, M. (1984). *Die Ästhetik als Antwort auf das Kopernikanische Weltbild*. Olms Verlag.
10. Pauli, W. "Der Einfluß archetypischer Vorstellungen auf die Bildung naturwissenschaftlicher Theorien bei Kepler" (1952) *Naturerklärung und Psyche*. Zurich: Rascher Verlag.
11. Hutcheson, F. (1973). *An inquiry concerning beauty, order, harmony, design*. Den Haag.
12. Planck, M. (1969). *Vorträge und Erinnerungen* (pp. 28, 65). Darmstadt: Wissenschaftlichen Buchgesellschaft.
13. Ibid.
14. The quotes from Leibniz are found in the essay, "Die mathematisch-physikalische Schönheit bei Leibniz" by Breger, H. (1994). *Revue internationale de philosophie*, *48*, 127–140.
15. Ibid.

Chapter 3

1. Hardy, G. H. (1967). *A mathematician's apology*. Cambridge: Cambridge University Press.
2. The efforts of psychologists concerning "Die Wissenschaft vom Schönen" are described by M. Plauen in an essay of the same title, which appeared in 1995 in the *Zeitschrift für Philosophische Forschung*, *49*, 54–75.
3. Ibid.
4. The golden section is discussed in the essay "Der Goldene Schnitt in der Natur—Harmonische Proportionen und die Evolution" by Richter, P. H., & Scholz, H.-J. It appeared in Küppers, B. O. (1987). *Ordnung aus dem Chaos* (pp. 175–214). Munich: Piper.
5. Pauli, W. "Der Einfluß archetypischer Vorstellungen auf die Bildung naturwissenschaftlicher Theorien bei Kepler" (1952) *Naturerklärung und Psyche*. Zurich: Rascher Verlag.
6. Ibid.
7. The golden section is discussed in the essay "Der Goldene Schnitt in der Natur—Harmonische Proportionen und die Evolution" by Richter, P. H., and Scholz, H.-J. It appeared in Küppers, B. O. (1987). *Ordnung aus dem Chaos* (pp. 175–214). Munich: Piper.
8. Pauli, W. "Der Einfluß archetypischer Vorstellungen auf die Bildung naturwissenschaftlicher Theorien bei Kepler" (1952) *Naturerklärung und Psyche*. Zurich: Rascher Verlag.
9. Benoît Mandelbrot (1987) wrote the large book, *Die fraktale Geometrie der Natur*. Basel: Birkhäuser.

Chapter 4

1. The quote by P. Friedländer is found in Curtin, D. W. (1982). *The aesthetic dimension of science*. New York.
2. The quotes by Einstein are found in (1993) *Gesammelten Schriften* (Collected papers), Vol. 5, pp. 271 and 604. Princeton: Princeton University Press.

3. Werner Heisenberg's autobiography was published in 1969 with the title *Der Teil und das Ganze* (Munich). The original work appeared in 1925 with the title, *Über quantentheoretische Umdeutung kinematischer und mechanischer Beziehungen*, Zeitschrift für Physik, *33*, 879–893.
4. Ibid.
5. Ibid.
6. The source of G. T. Fechner is from the volume edited by Florey E., and Briedback, O. (1993). *Das Gehirn—Organ der Seele?*, p. 269ff. Berlin.
7. Kekulé's dream is described in many places, for example, in C. G. Jung's book quoted in the foreword, *Der Mensch und seine Symbole*, pp. 37–38.
8. Musil, R. (1952). *Der Mann ohne Eigenschaften*. Hamburg: Rowohlt.
9. More on the alchemistic meditations and the archetypal speculation are found in my book, Fischer, E. P. (1995). *Die aufschimmernde Nachtseite der Wissenschaft*. Lengwil: Libelle Verlag. The subject discussed here is from the correspondence between Wolfgang Pauli and C. G. Jung published in 1991 by Springer Verlag. *Der Pauli-Jung-Dialog und seine Bedeutung für die moderne Wissenschaft* was the theme of a conference whose papers in 1995 in a book of the same name were published and edited by H. Atmanspacher et al. Berlin: Springer Verlag.
10. Eddington as quoted in A. Miller, *Insights of genius*. New York.
11. Hadamard, J. (1945). *Essay on the psychology of invention in the mathematical field*. Princeton: Princeton University Press.
12. More on Niels Bohr can be found in the biography I wrote about him, Fischer, E. P. (1987). *Die Lektion der Atome*. Munich: Piper.
13. Paul Dirac's remark and others on Einstein are found in my book, Fischer, E. P. (1996). *Einstein—Ein Genie und sein überfordertes Publikum*. Heidelberg: Springer Verlag.
14. Rutherford as quoted in Miller, A. (1996). *Insights of genius*.
15. The quotes by Einstein are found in (1993) *Gesammelten Schriften* (Collected papers), Vol. 5, pp. 271 and 604. Princeton: Princeton University Press.

Chapter 5

1. Simone Weil as quoted in Quadbeck-Seeger, H.-J. (Ed., 1988). *Zwischen den Zeichen* (p. 134). Weinheim.
2. Cohen, I. B. (1994) discussed revolution in general in *Revolutionen in der Naturwissenschaft*. Frankfurt: Suhrkamp.
3. The Bohr anecdotes are found for example in Werner Heisenberg's (1969) autobiography *Der Teil und das Ganze*. Munich.
4. Popper, K. (1969). *Logik der Forschung*. Tübingen: Mohr Verlag. In English translation (1965): *Logic of scientific discovery*. New York: Harper & Row.
5. The quotes by Einstein are found in (1993) *Gesammelten Schriften* (Collected papers), Vol. 5, pp. 271 and 604. Princeton: Princeton University Press.
6. The analysis of Robert Millikan is found in the essay, "On the art of scientific imagination" by Gerald Holton (1996, spring); it appeared in the journal *Daedalus*, 183–208.

7. Thomas Kuhn's (1970) thesis is found in his book *Structure of scientific revolutions*, Chicago: University of Chicago Press, in which he also briefly addresses the role of perception in science. He establishes his paradigm in the volume of essays *The essential tension* (Chicago: University of Chicago Press 1977).

8. Details on Lavoisier can be found in Serres, M. (Ed., 1994). *Elemente einer Geschichte der Wissenschaft*. Frankfurt: Suhrkamp.

9. Kuhn, T. (1970). *Structure of scientific revolutions*. Chicago: University of Chicago Press.

10. Fitzek, H., and Salber, W. (1996). *Gestalt Psychologie*. Darmstadt and Metzger, W. (1968) in the journal $n+m$, B. Mannheim (Ed.), *23*, 3–24 both discuss gestalt perception.

11. Lorenz, K. (1965). *Über tierisches und menschliches Verhalten*, in *Gesammelten Abhandlungen*. Munich: Piper.

12. See Gleick, J. (1992). *Genius*. New York: Pantheon Books, for more on Richard Feynman.

13. McAllister, J. (1996). *Beauty and the revolution in science*. Ithaca: Dordrecht.

Chapter 6

1. There are many books on symmetry, including those by Genz, H. (1988), *Symmetrie—Bauplan der Natur*. Munich: Piper. Mainzer, K. (1987). *Symmetrie der Natur*. Berlin and Zee, A. (1993). *Magische Symmetrie*. Frankfurt: Insel. The classic work on this subject is by Weyl, H. (1952). Symmetry. Princeton: Princeton University Press.

2. Kepler, J. (1987). *Vom sechseckigen Schnee*. Leipzig: Ostwalds Klassiker.

3. Landau, L. D. and Lifshitz, I. M. (1972–1990). *Course of theoretical physics*. Elmsford: Pergamon Press.

4. The work on symmetry of faces originates Perret, D. I., et al. (1994), appearing in the journal *Nature, 368*, 239–242.

5. Ibid.

6. Symons, D. (1979). *The Evolution of Human Sexuality*. Oxford: Oxford University Press.

7. Many biological aspects on aesthetics are discussed in Rentschler, I., Herzberg, B., & Epstein, D. (Eds., 1988). *Beauty and the brain*. Basel: Birkhauser Verlag (see suggested readings).

8. There are many books on symmetry, including those by Genz, H. (1988), *Symmetrie—Bauplan der Natur*. Munich: Piper. Mainzer, K. (1987). *Symmetrie der Natur*. Berlin and Zee, A. (1993). *Magische Symmetrie*. Frankfurt: Insel. The classic work on this subject is by Weyl, H. (1952). *Symmetry*. Princeton: Princeton University Press.

9. Ibid.

10. Gaffron, M. (1950) discusses looking right–left at pictures in "Right and left in pictures" published in *Art Quarterly, 13*, 312–331. For a description of the functional asymmetries of the brain, see Springer, S., & Deutsch, G. (1987), *Linkes-rechtes Gehirn*. Heidelberg.

11. Wölfflin, H. (1941). *Über das Rechts und Links im Bilde*. In: Gedanken zur Kunstgeschichte. Basel: Schwabe, pp. 82–90.
12. The important findings by Grüsser, O. J., et al. with the title "Cerebral lateralization and some implications for art, aesthetic perception, and artistic creativity" is found on pages 257–295 of the volume, *Beauty and the brain*.
13. Ibid.

Chapter 7

1. The quote by Edmund Burke can be found in Phillips, A. (Ed., 1990). *A philosophical enquiry into the origin of our ideas of the sublime and beautiful* (part 3, sec. 1, p. 83). New York: Oxford University Press.
2. The quotes from J. Brodsky come from the volume of essays *On grief and reason* (New York: Farrar, Straus, and Giroux 1995) and the volume *Das sichtbar Unsichtbare* appearing in 1994 in edition tertium (without editor).
3. Hilde Neunhöffer's views are found in her book, *Freie Frauen and ihre Rolle bei der Evolution des Homo sapiens* (Hamburg 1995) and in a valuable article with the title, "Ein Kind von Liebe und Freiheit," that appeared in 1997 in the series *Hörsaal Holzen* (Vol. 2).
4. Matt Ridley narrates the natural history of sexuality in his 1995 book *The red queen: Sex & the evolution of human nature* (New York: Viking Penguin).
5. Enquist, M., and Arak, A. (1994). "Symmetry, beauty and evolution." *Nature*, 372 169–172.
6. On the connection between sex and the symmetrical body, see the article by the same name, by Concar, D. (1995, April). *New Scientist, 22*, 40–44.
7. The biology of beauty. (1996, June 3). *Newsweek*, 43–50.
8. Richard Alexander as quoted by Ridley, M. (1995). *The red queen: Sex & the evolution of human nature* (p. 330). New York: Viking Penguin.
9. Geoffrey Miller as quoted by Ridley, M. (1995). *The red queen: Sex & the evolution of human nature* (p. 338). New York: Viking Penguin. The thesis by G. Miller is also discussed in Mestel, R. (1995). "Arts of seduction." *New Scientist, 23*, 26–31.
10. Dissanayake, E. (1995). *Homo aestheticus*. Seattle: University of Washington Press.
11. Hartmann, N. (1966). *Ästhetik*. Berlin.
12. The quotes from J. Brodsky come from the volume of essays *On grief and reason* (New York 1995) and the volume *Das sichtbar Unsichtbare* appearing in 1994 in edition tertium (without editor).

Chapter 8

1. The quote and the lecture by Adolf Portmann are found for example in the volume *Biologie und Geist* that appeared in 1956 in paperback in the Herder-Bücherei.
2. My views on the subject *Wertvolle Wissenschaft* are already summarized in the series of *Oldenburger Universitätsreden* (Number 87, 1997).

3. The essay by Konrad Lorenz is found in his *Gesammetlen Abhandlungen*, "Über tierisches und menschliches Verhalten," (Piper 1965).
4. Irvin Rock's book *Perception* (New York 1995) is the best work on perception.
5. The quote and the lecture by Adolf Portmann are found for example in the volume *Biologie und Geist* that appeared in 1956 in paperback in the Herder-Bücherei.
6. Ibid.
7. Fischer, E. P. (1995). *Die aufschimmernde Nachtseite der Wissenschaft*. Lengwil: Libelle Verlag.
8. The quotes by Wolfgang Pauli are found in the volumes that contain his *Wissenschaftlichen Briefwechsel*; they appeared in 1974, Heidelberg: Springer Verlag. Further references originate in the correspondence between W. Pauli and C. G. Jung that appeared in 1992.
9. More on Johannes Kepler and Richard Feynman can be found by the interested reader in Fischer, E. P. (1995). *Aristoteles, Einstein & Co*. Munich: Piper Verlag. The quotes from the works by Kepler are found in the long essay by Wolfgang Pauli, "Der Einfluß archetypischer Vorstellungen auf die Bildung naturwissenschaftlicher Theorien bei Kepler" in the book coauthored by Jung, C. G. (1952) *Naturerklärung und Psyche*. Zurich: Rascher Verlag, which contains precise references to *Mysterium cosmographicum, ad vitellionem paralipomena* and *Harmonices mundi*.
10. The quotes by Wolfgang Pauli are found in the volumes that contain his *Wissenschaftlichen Briefwechsel*; they appeared in 1974, Heidelberg: Springer Verlag. Further references originate in the correspondence between W. Pauli and C. G. Jung that appeared in 1992.
11. The text quoted by Michael Hauskeller is found in the volume *Hörsaal Holzen* (Vol. 2) appearing under the title "Verantwortliches Handeln" in 1997.
12. Ibid.
13. Jonas, H. (1984). *Das Prinzip Verantwortung*. Frankfurt: Suhrkamp Taschenbuch.
14. Milgram, S. (1974). *Obedience to authority: An experimental view*. New York: Harper & Row.
15. The journal *Nature* reports on the cloned sheep in Vol. 385 (1997) starting on p. 810. Additional commentary can be found in Vol. 386 of the same journal on p. 119. My analysis can be found in the March 6, 1997, issue of *Weltwoche* with the title, "Wer hat Angst vor meinem Klon."

Epilogue

1. The quote is from a conversation with Peter Sitte and he has Joseph Haydn to thank for it.
2. *The notebooks of Raymond Chandler* was published in 1976 in New York by Frank MacShane.
3. *The end of science* by John Horgan was published in New York. *The nature of space and time* by Hawking and Penrose appeared in 1996 with Princeton.
4. Jonas, H. (1984). *Das Prinzip Verantwortung* (p. 392).

5. Hartmann, N. (1966). *Ästhetik* (p. 48).
6. The quotes by Joseph Brodsky are found in an essay in the 1994 volume appearing in the edition tertium (without editor) *Das sichtbar Unsichtbare*.
7. Brodsky, J. (1995). *On grief and reason*. New York: Farrar, Straus, and Giroux, p. 205.
8. The newly founded Hochschule Holzen (Holzen Institute) summarized its purpose in a "Holzener Manifest" that can be primarily attributed to Martin Rabe and composed together with Daniel Meynen and the author of this book. The manifest can be ordered from Holzen at the following address: Hochschule Holzen, Kirchstr. 8, D 79400 Kandern-Holzen; tel: +49-76 26-9 15 80 and fax: +49-76 26-91 58 23.

Suggested Reading

Chandrasekhar, S. (1987). *Truth and beauty*. Chicago: University of Chicago Press.
Cooper, D. (Ed., 1995). *A companion to aesthetics*. Oxford: Blackwell Publishers.
Cramer, F., & Kaempfer, W. (1992). *Die Natur der Schönheit*. Frankfurt: Insel.
Curtin, D. W. (Ed., 1982). *The aesthetic dimension of science*. New York.
Dissanayake, E. (1995). *Homo aestheticus*. Seattle: University of Washington Press.
Gregory, R., et al. (Ed., 1995). *The artful eye*. Oxford: Oxford University Press.
Hauskeller, M. (Ed., 1994). *Was das Schöne sei*. Munich.
Hutcheson, F. (1973). *An inquiry concerning beauty, order, harmony, design*. The Netherlands.
Löw, R. (1994). *Über das Schöne*. Stuttgart: Weitbrecht.
McAllister, J. (1996). *Beauty and revolution in science*. Ithaca: Cornell University Press.
Miller, A. (1996). *Insights of genius*. New York: Copernicus Books.
Portmann, A. (1973). *Vom lebendigen*. Frankfurt.
Rentschler, I., Herzberger, B., & Epstein, D. (Eds., 1988). *Beauty and the brain*. Basel: Birkhauser Verlag.
Ridley, M. (1995). *The red queen*. New York: Viking Penguin.
Rock, I. (1995). *Perception*. New York.
Santayana, G. (1955). *The sense of beauty*. New York: Dover Books.
Schweizer, H. R. (1967). *Vom ursprünglichen Sinn der Ästhetik*. Zug.
Seel, M. (1991). *Eine Ästhetik der Natur*. Frankfurt: Suhrkamp.
Tauber, A. (Ed., 1996). *The elusive synthesis—Aesthetics and science*. Dordrecht: Kluwer Academic Publishers.
Turner, F. (1990). *Beauty—The value of values*. Charlottesville: University Press of Virginia.

Index